Energy Dissipation in Hydraulic Structures

IAHR Monograph

Series editor

Peter A. Davies
Department of Civil Engineering,
The University of Dundee,
Dundee,
United Kingdom

The International Association for Hydro-Environment Engineering and Research (IAHR), founded in 1935, is a worldwide independent organisation of engineers and water specialists working in fields related to hydraulics and its practical application. Activities range from river and maritime hydraulics to water resources development and eco-hydraulics, through to ice engineering, hydroinformatics and continuing education and training. IAHR stimulates and promotes both research and its application, and, by doing so, strives to contribute to sustainable development, the optimisation of world water resources management and industrial flow processes. IAHR accomplishes its goals by a wide variety of member activities including: the establishment of working groups, congresses, specialty conferences, workshops, short courses; the commissioning and publication of journals, monographs and edited conference proceedings; involvement in international programmes such as UNESCO, WMO, IDNDR, GWP, ICSU, The World Water Forum; and by co-operation with other water-related (inter)national organisations. www.iahr.org

Energy Dissipation in Hydraulic Structures

Hubert Chanson

*School of Civil Engineering, University of Queensland,
Brisbane, Australia*

CRC Press
Taylor & Francis Group
Boca Raton London New York

CRC Press is an imprint of the
Taylor & Francis Group, an **informa** business

A BALKEMA BOOK

Cover photo credit: Gold Creek dam stepped spillway in operation on 23 February 2015 (Photograph H. Chanson)

Published by:
CRC Press/Balkema
P.O. Box 447, 2300 AK Leiden, The Netherlands
e-mail: Pub.NL@taylorandfrancis.com
www.crcpress.com – www.taylorandfrancis.com

First issued in paperback 2020

ISBN 13: 978-0-367-57573-1 (pbk)
ISBN 13: 978-1-138-02755-8 (hbk)

Visit the Taylor & Francis Web site at
http://www.taylorandfrancis.com

and the CRC Press Web site at
http://www.crcpress.com

Typeset by V Publishing Solutions Pvt Ltd., Chennai, India

Library of Congress Cataloging-in-Publication Data

Chanson, Hubert.
  Energy dissipation in hydraulic structures / Hubert Chanson.
    p. cm. – (IAHR monograph)
  Includes bibliographical references and index.
  ISBN 978-1-138-02755-8 (hardcover : alk. paper) – ISBN 978-1-315-68029-3
  (ebook : alk. paper) 1. Hydrodynamics. 2. Hydraulic structures. 3. Energy
  dissipation. 4. Diversion structures (Hydraulic engineering) I. Title.
  TC171.C473 2015
  627–dc23

                                                            2015011109

About the IAHR Book Series

An important function of any large international organisation representing the research, educational and practical components of its wide and varied membership is to disseminate the best elements of its discipline through learned works, specialised research publications and timely reviews. IAHR is particularly well-served in this regard by its flagship journals and by the extensive and wide body of substantive historical and reflective books that have been published through its auspices over the years. The IAHR Book Series is an initiative of IAHR, in partnership with CRC Press/ Balkema – Taylor & Francis Group, aimed at presenting the state-of-the-art in themes relating to all areas of hydro-environment engineering and research.

The Book Series will assist researchers and professionals working in research and practice by bridging the knowledge gap and by improving knowledge transfer among groups involved in research, education and development. This Book Series includes Design Manuals and Monographs. The Design Manuals contain practical works, theory applied to practice based on multi-authors' work; the Monographs cover reference works, theoretical and state of the art works.

The first and one of the most successful IAHR publications was the influential book "*Turbulence Models and their Application in Hydraulics*" byW. Rodi, first published in 1984 by Balkema. I. Nezu's book "*Turbulence in Open Channel Flows*", also published by Balkema (in 1993), had an important impact on the field and, during the period 2000–2010, further authoritative texts (published directly by IAHR) included *Fluvial Hydraulics* by S. Yalin and A. Da Silva and *Hydraulicians in Europe* by W. Hager. All of these publications continue to strengthen the reach of IAHR and to serve as important intellectual reference points for the Association.

Since 2011, the Book Series is once again a partnership between CRC Press/ Balkema – Taylor & Francis Group and the Technical Committees of IAHR and I look forward to helping bring to the global hydro-environment engineering and research community an exciting set of reference books that showcase the expertise within IAHR.

Peter A. Davies
University of Dundee, UK
(Series Editor)

Table of contents

Preface

Recent advances in technology have permitted the construction of large dams, reservoirs and channels. This progress has necessitated the development of new design and construction techniques, particularly with the provision of adequate flood release facilities. Chutes and spillways are designed to spill large water discharges over a hydraulic structure (e.g. dam, weir) without major damage to the structure itself and to its environment. At the hydraulic structure, the flood waters rush as an open channel flow or free-falling jet, and it is essential to dissipate a very significant part of the flow kinetic energy to avoid damage to the hydraulic structure and its surroundings. Energy dissipation may be realised by a wide range of design techniques. A number of modern developments have demonstrated that such energy dissipation may be achieved (a) along the chute, (b) in a downstream energy dissipator, or (c) a combination of both.

The magnitude of turbulent energy that must be dissipated in hydraulic structures is enormous even in small rural and urban structures. For a small storm waterway discharging 4 m³/s at a 3 m high drop, the turbulent kinetic energy flux per unit time is 120 kW! At a large dam, the rate of energy dissipation can exceed tens to hundreds of gigawatts; that is, many times the energy production rate of nuclear power plants. Many engineers have never been exposed to the complexity of energy dissipator designs, to the physical processes taking place and to the structural challenges. Several energy dissipators, spillways and storm waterways failed because of poor engineering design. It is believed that a major issue affecting these failures was the lack of understanding of the basic turbulent dissipation processes and of the interactions between free-surface aeration and flow turbulence.

In that context, an authoritative reference book on energy dissipation in hydraulic structures is proposed here. The book contents encompass a range of design techniques including block ramps, stepped spillways, hydraulic jump stilling basins, ski jumps and impact dissipators.

Chapter 1

Introduction: Energy dissipators in hydraulic structures

H. Chanson
School of Civil Engineering, The University of Queensland,
Brisbane, QLD, Australia

ABSTRACT

Many regions on our planet are subjected to hydrological extremes. Dams and reservoirs constitute a very efficient means to provide both long-term water reserves and flood protection. During major rainfall events, large water inflows into the reservoir induce a rise in the reservoir level, and a spillway system must be installed to spill safely the flood waters. Two key challenges during the spillway design are conveyance and energy dissipation. Energy dissipation on dam spillways can be achieved by a range of dissipator designs. A massive challenge is the magnitude of the rate of energy dissipation in the spillway structure during major floods.

1.1 PRESENTATION

The dependence of life on water is absolute on Earth. Plants, animals and humans depend all upon the availability of water to sustain life. Water constitutes the primary constituent of the oceans, rivers and subterraneous fluids on our planet. In its various forms, water occupies about 1.4×10^{18} m³ on Earth (Berner and Berner 1987). The water cycle on the planet includes the evaporation of water, atmospheric circulation of vapour, precipitations and flowing streams both above and below ground. Through this cycle, air and water constantly interact: in the atmosphere, at the sea surfaces and on the mainland. These continuous exchanges are most important for the biological and chemical equilibrium of the planet including the balance between nitrogen, oxygen and carbon dioxide in the atmosphere.

Despite the continuous water cycle, the rainfall pattern on our planet is not uniform. Many regions are subjected to hydrological extremes with high spatial and temporal variability. On one hand, extreme rainfalls can be experienced: e.g., 3,721 mm of rain in 4 days at Cherrapunji (India) on 12–15 September 1974 constitutes a world's record for that duration (WMO 1994). Intense rainfalls convert into some surface runoff down gullies and brooks, converging into valleys and streams to generate large river floods. Recent catastrophes included the inundations of Vaison-la-Romaine (France) in 1992, Nîmes (France) in 1998, Rockhampton, Bundaberg, and Brisbane, Queensland (Australia) during the 2010–2011 summer and again in summer 2013, the flood of the Mississippi River (USA) in spring 2011, the Thames Valley (UK) in winter 2014. IAHS (1984) presented a catalogue of the world's maximum floods highlighting astonishing record flow rates, although most maximum discharge estimates are of poor precision with no or little information on measurement accuracy. On the other hand, many countries can experience hot and dry periods including long droughts. As an illustration, Australia, a dry continent, has experienced recurrent long droughts for the last 250 years: e.g., the longest drought (1958–1967) in arid Central Australia, the WWII drought (1939–45) during which 30 millions sheep died, the Federation drought (1900s) during which the Murray River stopped flowing for 6 months and the country's sheep population dropped from 106 millions down to 54 millions.

Dams and reservoirs constitute one of the most efficient means to provide our community with long-term water reserves and flood protection. Although some projects are built for flood mitigation or water supply only (e.g. Grande Dixence, Switzerland), most reservoirs are multi-purpose encompassing both water supply and flood mitigation. A dam is basically a man-made hydraulic structure built across a valley to provide water storage (Fig. 1.1). During major rainfall events, large water inflows into the reservoir induce a rise in the reservoir level, with the risk of dam overtopping during extreme floods. A spillway system must be installed: that is, an aperture designed to spill safely the flood waters above, below or besides the dam wall (Fig. 1.2).

Recent advances in technology have permitted the construction of large man-made dams and reservoirs. These progresses have necessitated the development of new designs and construction techniques, particularly for the provision of adequate flood release facilities. The spillways must be designed to spill large water discharges over a hydraulic structure without major damage to the structure itself and to its environment.

1.2   SPILLWAYS AND ENERGY DISSIPATORS

1.2.1   Basic considerations

The spillway system is designed to pass safely flood waters above, below, within or around the dam. A safe spillway operation is critical. Many dam failures were caused by poor spillway design and insufficient spillway discharge capacity (Smith 1971, Lempérière 1993). The latter is particularly critical in the case of earthfill and rockfill embankments, while concrete dams are more likely to withstand some moderate overtopping. Most dam design and construction require some conservatism, because a failure may cause a serious hazard to human life.

A large majority of dams are equipped with an overflow spillway system which includes typically a crest, a chute and an energy dissipator at the downstream end. The preliminary design stages include the selection of the maximum outflow discharge and the type of spillway (USBR 1965, Novak et al. 2001). The spillway capacity is derived from hydrological studies (Probable Maximum Precipitation, Probable Maximum Flood), together with Maximum Observed Floods (MOF). The design discharge is selected as a function of the surcharge storage capacity and maximum reservoir inflow. The type of spillway is a function of a number of parameters, including the flood hydrograph characteristics, damage to upstream and downstream catchments, damage to the dam and spillway in case of failure, and the possible combined use of outlet(s) and spillway(s).

The two key challenges during the spillway design process are conveyance and energy dissipation. Conveyance means the (safe) transport of large amount of waters

Figure 1.1 Old hydraulic structures. (A) Caromb dam or *Barrage du Paty* (France) in December 1994, dam completed during the 18th century. (B) Les Peirou dam (France) in June 1998. Completed in 1891, the 19 m high arch dam was built on the Glanum Roman dam foundations. (C) Urft dam (Germany) in February 2013. Completed in 1905, the 55 m high dam is equipped with a large stepped cascade.

Figure 1.2 Spillway systems. (A) Barrage Mercier (Canada) and spillway operation on 14 July 2002. (B) Wivenhoe dam spillway (Australia) on 21 December 2010. (C) Miyagase dam (Japan) on 3 October 2012. Chute spillway with mid-level outlets in operation. (D) Fukashiro dam (Japan) on 3 October 2012. Note the counterweir in the foreground right. (E) North Pine dam (Australia) on 22 May 2009.

Figure 1.3 Hydraulic jump stilling structure. (A) Shih Kang dam (Taiwan) on 12 Nov. 2010. (B) Jiji weir on Zhoushui River (Taiwan) on 11 January 2014.

into the intake, crest and wastewater channel (chute). The conveyance of the system is closely linked to the spillway crest and chute/channel design. It relies upon fundamental hydrodynamics and theoretical studies with a range of trusted and proven solutions. The energy dissipation takes place down the chute and at the downstream end of the spillway. The amount of kinetic energy to be dissipated can be gigantic (see below), and it must be dissipated safely before the floodwaters rejoin the natural river system. The design of the energy dissipator(s) relies upon some sound physical modelling combined with solid prototype experiences.

Energy dissipation on dam spillways is achieved usually by a standard stilling basin downstream of a steep chute in which a hydraulic jump takes place, converting the flow from supercritical to subcritical conditions (Fig. 1.3), a high velocity water jet taking off from a ski jump and impinging into a downstream plunge pool (Fig. 1.4), a plunge pool in which the chute flow impinges and the kinetic turbulent energy is dissipated in turbulent recirculation, or a stepped spillway chute followed by a small stilling structure (Fig. 1.5). A related design is a mid-level or bottom outlet discharging into a deep pool of water (Fig. 1.6). In any case, the energy dissipator must be designed to dissipate the excess in kinetic energy at the end of the spillway system before it re-joins the natural stream.

A massive challenge is the magnitude of the rate of energy dissipation in the spillway structure at design flow conditions. As an example, the Paradise Dam spillway system (Fig. 1.5B) experienced several major floods between 2010 and 2013. The dam is 37 m high and the observed peak discharge nearly reached 17,000 m³/s during that period, although well below the design flow (Fig. 1.7). At the peak flood flow, the spillway dissipated energy at a rate of 7.5 GW (or 7,500,000,000 W) comparable to and larger than the energy production rate of eight 900 MW nuclear reactors. For comparison, the combined power output of the two Three Miles Island nuclear reactors was 1.75 GW, while the power output of Fukushima nuclear power plant was 4.7 GW, and the Chernobyl nuclear power station produced 4 GW, prior to their respective accidents. The energy dissipation at a dam spillway can be enormous and the design of an energy dissipator is far from trivial.

Figure 1.4 Flip bucket spillway systems. (A) Clyde dam spillway (New Zealand) (Courtesy of Graham Quinn & Contact Energy). (B) Mudan dam spillway (Taiwan) on 21 November 2006.

Figure 1.5 Stepped spillways and downstream stilling structures. (A) Les Olivettes dam (France) in March 2003 (Courtesy of Mr and Mrs Jacques Chanson). (B) Paradise dam (Australia) on 5 March 2013.

1.2.2 Operational issues

The operational experience of dam, reservoirs and spillway systems showed that the water conveyance and safe energy dissipation remain some key challenges for the lifetime of the structure. Dam overtopping during construction is a well-known problem, with limited to major consequences depending upon the type of dam construction. Well-known operational challenges include cavitation, air entrainment, reservoir

Figure 1.6 Bottom outlets. Three Gorges Project (China) on 20 October 2004. Each outlet discharged 1700 m³/s with an exit velocity of 35 m/s.

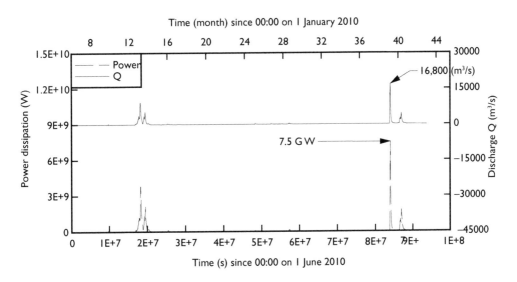

Figure 1.7 Paradise dam spillway operation in 2010–2013: spillway discharge and hydraulic power dissipated in the spillway system (Data courtesy of Sunwater, calculations by the author).

sedimentation, scour, debris blockage, mechanical failure, as well as human errors. A number of handbooks reviewed these problems (USBR 1965, Novak et al. 2001).

Cavitation is an implosive process inducing very high shear stresses and pressures, capable of destroying any surface (Knapp et al. 1970, Franc and Michel 2004). In hydraulic structures, cavitation damage can be observed in particular in high head systems and stilling basins. Micro-jet velocities of up to 300 m/s and pressures up to 1,500 MPa were recorded at the Dartmouth dam (Lesleighter 1983). Cavitation pitting can lead to massive scour as seen at Glen Canyon dam (Falvey 1990). Air entrainment, either by natural self-aeration or by means of aeration devices, may be used to prevent the risks of cavitation damage (Peterka 1953, Russell and Sheehan 1974). The flow aeration must however be controlled, because too much air may have adverse effects (flow bulking, blowback, supersaturation) in tunnels, conduits and channels. Reservoir sedimentation and scour at dam toe and river banks are well-known challenges to dam operators, illustrating our limited understanding of the interactions between large flood flows and sediments. The failure of hydraulic structures linked with debris blockage and mechanical failures have been relatively well-documented: e.g., the dam overtopping accidents of Palagnedra dam (Switzerland) in 1978 and at Pinet dam (France) in 1982 caused by debris blockage, the gate failures at Goulburn weir (Australia) in 1978, Folsom dam (USA) in 1995, and Tous dam (Spain), the latter leading to a complete dam failure in 1982. Human errors may encompass operation errors (e.g. Belci dam (Romania) failure in 1991), lack of maintenance and design errors. These also include conceptual errors like at the Malpasset dam (France).

The re-evaluation of spillway discharge capacity, including the spillway re-design, is a necessary challenge, especially in many regions with extreme hydrology and limited records, like Africa, Asia and Australia. For the last few decades, a number of dams sustained a flood significantly larger than the design flow, before their spillway capacity was re-evaluated (Lemperiere et al. 2012). A further number of structures experienced discharges larger than the design capacity and failed. Worldwide there is a lack of broad guidelines to estimate the maximum spillway capacity: "*consensus is far from being reached on the issue of spillways design*" capacity (Lemperiere et al. 2012).

1.3 STRUCTURE OF THE MONOGRAPH

Energy dissipation is closely associated to the spillway design and there are a number of different designs. This monograph reviews the key features of block ramp energy dissipation, stepped spillways and cascades, hydraulic jumps and stilling basins, ski jumps and flip buckets, and impact dissipators.

It is the aim of this monograph to regroup the state-of-the-art expertise and experience on energy dissipators for dam spillways and hydraulic structures. The chapters were prepared by world experts in spillway engineering and research, who are actively involved in the IAHR Hydraulic Structures Technical Committee.

ACKNOWLEDGEMENTS

The author thanks his students, former students, co-workers and all the people who provided relevant informations. The author acknowledges the financial support of the Australian Research Council (Grants DP0878922 & DP120100481).

REFERENCES

Berner, E.K., and Berner, R.A. (1987). *The Global Water Cycle: Geochemistry and Environment*. Prentice Hall, Englewood Cliffs NJ, USA.

Falvey, H.T. (1990). *Cavitation in Chutes and Spillways*. USBR Engineering Monograph, No. 42, Denver, Colorado, USA, 160 pages.

Franc, J.M., and Michel, J.M. (2004). *Fundamentals of Cavitation*. Springer, Series Fluid Mechanics and Its Applications, Vol. 76, 306 pages.

IAHS (1984). *World catalogue of maximum observed floods*. International Association of Hydrological Sciences, IAHS Publication No. 143, 354 pages.

Knapp, R.T., Daily, J.W., and Hammitt, F.G. (1970). *Cavitation*. McGraw-Hill Book Company, New York, USA.

Lempérière, F. (1993). Dams that have Failed by Flooding: an Analysis of 70 Failures. *International Water Power and Dam Construction*, Vol. 45, No. 9/10, pp. 19–24.

Lempérière, F., Vigny, J.P., and Deroo, L. (2012). New methods and criteria for designing spillways could reduce risks and costs significantly. Hydropower and dam construction, (3), 120–128.

Lesleighter, E. (1983). Cavitation in High-Head Gated Outlets – Prototype Measurements and Model Simulation. *Proceedings of 20th IAHR Biennial Congress*, Moscow, Vol. 3, Sec. B, pp. 495–503.

Novak, P., Moffat, A.I.B., Nalluri, C., and Narayanan, R. (2001). *Hydraulic Structures*. Spon Press, London, UK, 3rd edition, 666 pages.

Peterka, A.J. (1953). The Effect of Entrained Air on Cavitation Pitting. *IAHR/ASCE Joint Meeting Paper*, Minneapolis, Minnesota, 507–518.

Peterka, A.J. (1958). *Hydraulic Design of Stilling Basins and Energy Dissipators*. USBR Engineering Monograph, No. 25, Denver, Colorado, USA, 239 pages.

Russell, S.O., and Sheehan, G.J. (1974). Effect of Entrained Air on Cavitation Damage. *Canadian Journal of Civil Engineering*, 1, 97–107.

Smith, N. (1971). *A History of Dams*. The Chaucer Press, Peter Davies, London, UK.

USBR (1965). *Design of Small Dams*. US Department of the Interior, Bureau of Reclamation, Denver CO, USA, 1st edition, 3rd printing.

WMO (1994). *Guides to Hydrological Practices*. World Meteorological Organisation, WMO-No. 168, 5th edition, 735 pages.

Chapter 2

Energy dissipation at block ramps

S. Pagliara & M. Palermo
DESTEC – Department of Energy Engineering, Systems,
Land and Construction, University of Pisa, Pisa, Italy

ABSTRACT

Block ramps are hydraulic structures which are often used in practical applications both to control sediment transport and, at the same time, to assure a correct balance between hydraulic functioning and environmental impact. This structure typology has become more and more popular especially in the last century. The necessity to give eco-sustainable answers to hydraulic problems related to stream restoration has been considered essential, in the perspective of a correct balance between hydraulic structure performances and preservation of natural habitats. Block ramps could be considered one of the answers for the afore-mentioned scenario as they minimize their impact on the contexts in which they are located and offer several advantages in terms of both energy dissipation and sediment transport control. The construction techniques have been substantially improved during the last decades. In fact, this structure typology has been employed for wider and wider ranges of conditions and configurations. One of the main peculiarities of this approach is the capacity to dissipate a larger energy amount than other traditional transversal stream-restoration structures, such as check dams. Thus, significant efforts were spent by the scientific community in order to optimize their energy efficiency. The present chapter aims to synthetize the actual knowledge regarding dissipative processes associated with block ramps. In particular, a critical discussion of the main parameters influencing the dissipative process is proposed and the main achievements present in literature up to the date are reported, in order to furnish a comprehensive summary which can be useful for both researchers and hydraulic engineers.

2.1 INTRODUCTION

The conjugation of anthropic necessities and environmental care is one of the challenge of the science. It requires a big effort from the researcher community because of both the variety and complexity of scenarios. In particular, hydraulic engineers have a prominent role in the preservation and improvement of environmental quality and sustainability. Rivers, torrents, channels and, generally, water bodies require a special and increasing attention, because of the importance of water for humanity. Thus, hydraulic engineers are called to provide solutions for both water quality improvement and natural environment preservation, and, at the same time, they have to assure higher and higher performances of anthropic structures. In this perspective, stream restoration is one of the key features and most relevant challenge. Generally, two aspects should be carefully taken into consideration to design hydraulic structures: 1) the structure has to control sediment transport; and 2) it has to reduce flow energy. Traditional transversal concrete structures can give answers to both these aspects, but, generally, they constitute an isolated and completely external element respect to the in situ contexts. Block ramps have the advantage to furnish answers that can assure both hydraulic functioning and low-environment impact, thus they have become more and more popular. They are hydraulic structures made of loose or fixed rocks located on a sloped bed. Generally, in their base configuration, they are characterized by a steep slope bed, preceded and followed by a milder slope bed configuration. In fact, in usual applications, the flow is in sub-critical conditions both upstream and downstream of the structure and it is super-critical on the structure itself. The passage from the two flow regimes occurs in association with the ramp entrance, at which critical flow conditions are reached. The dissipative process occurs both on the ramp itself, because of the localized variations of stream slope and high relative roughness conditions due to the base material constituting the ramp, and eventually downstream of it. In the downstream stilling basin a hydraulic jump can eventually occur, thus contributing to dissipate a further amount of flow energy. According to the downstream hydraulic conditions, a hydraulic jump can also occur on the ramp itself, at different longitudinal locations. In this last case, the dissipative process is mainly localized on the ramp itself. Thus, the energy dissipation process can depend on several aspects, many of which have been explored by different researchers especially in the last decades. A picture of a block ramp is shown in Figure 2.1.

In particular, the first studies were mostly related to structural stability of the ramp. In fact, its hydraulic functioning can be assured only when the structure itself does not face problems of instability of the blocks which can eventually lead to a collapse of the ramp. In particular, some of the first studies related to block ramps were conducted by Platzer (1993). The author analysed the hydraulic behaviour of this type of hydraulic structure in the case in which its slope is 1V:10H. Further

Figure 2.1 Picture of a block ramp built in a river, Carpathian Mountains (Poland).

studies on stepped chutes, which exhibit a similar behaviour of block ramps in the case of skimming flow conditions, were conducted successively by, among others, Chamani and Rajaratnam (1999a,b), Diez-Cascon et al. (1991), Peyras et al. (1992), Stephenson (1991), Christodoulou (1993), Pagliara and Palermo (2011c) and Pagliara and Palermo (2013). In particular, Chanson (1994) conducted a detailed analysis on energy dissipation on stepped chutes in the case of both nappe and skimming flow and proposed useful relationships in order to estimate it. Understanding of the effect of flow resistance on flow characteristics in the presence of a rough sloped bed was deepened by Rice et al. (1998), Bathurst (1978) and Bathurst et al. (1981). In the last two mentioned studies, the effect of roughness was analysed and a methodology to classify the flow conditions on the rough beds was developed. Whittaker and Jäggi (1986), Robinson et al. (1997) and Pagliara and Palermo (2011a) gave substantial contributions in order to predict the hydraulic conditions and ramp configuration which can assure loose blocks stability on the ramp itself. They furnished experimental relationships by which one can obtain an estimation of the critical discharge according to both hydraulic conditions and ramp configurations. Only recently, a more comprehensive analysis of both scour and dissipative processes associated with block ramps has been proposed. In particular, at the University of Pisa, detailed experimental studies have been conducted in the last decade, by which it is possible to furnish a complete and global approach to block ramp design. The analysis of energy dissipation on both base and reinforced block ramps was conducted by Pagliara and Chiavaccini (2006a,b), respectively. The authors analysed the behavior of block ramps in terms of energy dissipation varying both bed roughness and ramp slope. In particular, they furnished an interpretation of hydraulic phenomena occurring on block ramps in the presence of a downstream horizontal smooth bed in which a hydraulic jump took place. Successively, the analysis was focused on the entire dissipative process, including also the energy dissipated in the stilling basin in the case of a mobile bed. In fact, an analysis of the erosive phenomenon downstream of block ramp in a prismatic channel (ramp width equal to stilling basin width) conducted by Pagliara and Palermo (2008a,b), Pagliara and Palermo (2010a) and Pagliara et al. (2012) provided evidence to show that the role of the hydraulic jump associated with a block ramp is essential in order to understand both the hydraulic characteristics and flow pattern. Thus, Pagliara et al. (2008) conducted a series of experimental tests for different hydraulic jump locations (i.e., for both submerged and un-submerged ramp configurations), extending the previous findings of Pagliara and Chiavaccini (2006a), i.e., taking into consideration the global dissipative process from the ramp entrance to the end of the hydraulic jump. This last study specified and clarified the effect of the hydraulic jump on the global dissipative process. However, a further analysis was also conducted in the presence of protection structures in the stilling basin by Pagliara and Palermo (2010b), who analysed the global dissipative process when rock sills are located downstream of the ramp in different spatial positions. These studies were all conducted using a prismatic channel configuration, as specified above. But successive analyses conducted by Pagliara et al. (2009a), Pagliara and Palermo (2011b), Pagliara et al. (2011) in the case in which the stilling basin width was larger than that of the ramp, proved that both the hydrodynamics and scour processes occurring are subjected to huge and significant variations. Namely, the hydraulic jump occurring in an abrupt enlarged stilling basin is characterized by peculiar flow characteristics.

First of all, a 3D hydraulic jump takes place and its flow structure depends on both the ratio between stilling basin and ramp width and hydraulic conditions (discharge and tailwater level). Based on these observations, Pagliara et al. (2009b) analysed the global dissipative phenomenon in the presence of a block ramp with a downstream enlarged movable stilling basin, extending the previous findings to a more general configuration. Finally, Pagliara and Palermo (2012) further analysed the effect of stilling basin geometry on the dissipative process in the case of an asymmetric enlargement. The analysis of the energy dissipation was also conducted in the presence of sediment transport occurring on the block ramp (see Pagliara et al., 2009c), by which it was possible to state that, in this condition, there is a partial energy recovery. Other authors have also conducted experimental tests related to block ramps and energy dissipation phenomenon. Namely, Oertel and Schlenkhoff (2012) conducted a series of experimental tests in the presence of crossbar block ramps. In particular, they proposed a comprehensive analysis focusing on both hydraulic and geometric parameters affecting the dissipative process. A further study was proposed by Ahmad et al. (2009), in which the effect of staggered boulders located on block ramps was analysed in terms of energy dissipation. Despite the conspicuous literature dealing with dissipative process associated with block ramps, several aspects and mechanisms are still not completely clear. Thus, the present chapter aims both to furnish a general and critical picture of the actual knowledge regarding the dissipative processes occurring in the presence of block ramps and to stimulate new researches on this topic.

2.2 ENERGY DISSIPATION ON BASE BLOCK RAMPS

The base configuration of a block ramp was analysed by Pagliara and Chiavaccini (2006a) in terms of energy dissipation. It consisted in a sloped rough bed, on which rock elements were glued, followed by a horizontal smooth stilling basin, having the same width of the ramp itself. The authors conducted their experimental tests, for which a free hydraulic jump was occurring entirely in the stilling basin, varying the ramp slope S in the range $0.08 < S < 0.25$. They tested several ramp materials in order to analyse the effect of roughness on the dissipative process. Namely, they conducted experiments using five different ramp bed materials whose average diameter d_{50} varied between 1 mm to 88 mm. In particular, they estimated the relative energy loss ΔE_r due to both geometric configuration and hydraulic characteristics. In Figure 2.2, the experimental apparatus modelled by Pagliara and Chiavaccini (2006a) is sketched, along with a picture showing the base ramp.

In this figure, k is the critical depth occurring at the ramp entrance, h_1 is the flow depth downstream of the ramp toe, h_2 the flow depth downstream of the hydraulic jump and H is the ramp height. Pagliara and Chiavaccini (2006a) directly estimated both the total upstream energy $E_0 = H + 1.5k$ and the downstream energy at the toe of the structure $E_1 = h_1 + V_1^2/(2g)$, by measuring both the flow discharge and the water depth h_1. Applying the Buckingham Π theorem, the authors were able to furnish a non-dimensional functional relationship which identifies and specifies the dependence of the relative non-dimensional energy loss $\Delta E_r = \Delta E_1/E_0$, where $\Delta E_1 = (E_0 - E_1)$, on the following variables:

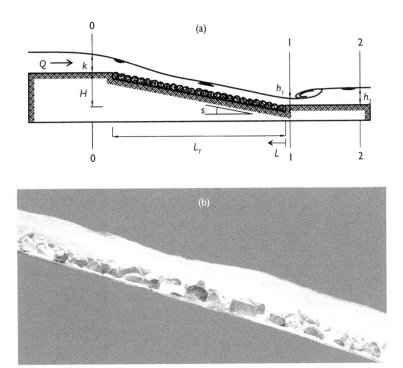

Figure 2.2 (a) Definition sketch of the experimental apparatus along with the indication of the main geometric and hydraulic parameters; (b) picture of a base ramp configuration.

$$\Delta E_r = f\left(\frac{k}{d_{50}}, \frac{k}{H}, S, \text{Re}, \text{We}, C_{mean}, \frac{b}{k}\right) \tag{2.1}$$

In which k is the critical depth, b is the ramp width (equal to the stilling basin width), H is the ramp height, S the ramp slope, Re the Reynolds number, We the Weber number, and C_{mean} the non-dimensional mean air concentration of the flow on the structure. For tested ramp slopes, Pagliara and Chiavaccini (2006a) proved that both the effects of C_{mean} and We on the relative energy loss are negligible, thus the dependence on these two variables can not be considered in order to obtain a predicting equation. In addition, they also proved that both b/k and Re (Lawrence, 1997, Ferro, 1999) do not affect significantly the relative energy loss. Based on these observations, authors concluded that the relative energy loss ΔE_r can be expressed by the following functional relationship:

$$\Delta E_r = f\left(\frac{k}{d_{50}}, \frac{k}{H}, S\right) \tag{2.2}$$

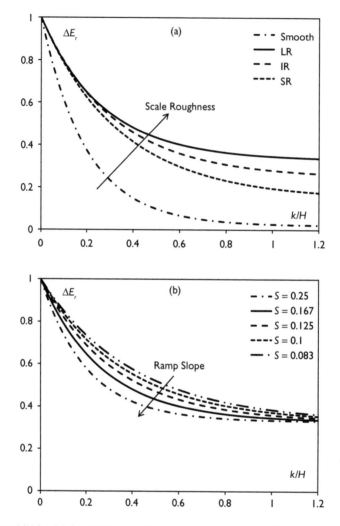

Figure 2.3 ΔE_r vs k/H for (a) $S = 0.167$ and different scale roughness and for (b) large scale roughness and different ramp slopes.

It means that ΔE_r only depends on relative roughness k/d_{50}, relative height k/H and bed slope S. The effects of the proposed independent parameters were analysed. In particular, based on Bathurst (1981), the authors provided a new classification, by which one can distinguish the roles of roughness conditions. Namely, block ramps are short structures for which uniform flow conditions are not reached in general; thus, it was necessary to furnish a classification of relative roughness which should be independent of the flow regime occurring on the ramp itself. Pagliara and Chiavaccini (2006a) stated that large scale roughness (LR) occurs for $k/d_{50} < 2.5$, intermediate scale roughness (IR) for $2.5 < k/d_{50} < 6.6$ and small scale roughness (SR) for $6.6 < k/d_{50} < 42$. Based on these deductions, the authors proposed graphs $\Delta E_r(k/H)$ by which it was possible to illustrate both the effects of ramp slope and relative roughness on the dissipative process. In Figure 2.3a–b, two parametric graphs are reported. Figure 2.3a is relative to a block ramp configuration whose slope is $S = 0.167$ and for different relative roughness conditions, including smooth ramp. It can be easily noted that, for constant ramp slope, the relative energy dissipation increases passing from smooth to large relative roughness conditions (LR). In addition, the ramp slope also has a significant effect on the dissipative process. In fact, Figure 2.3b shows the effect of S on ΔE_r, proving that, for constant scale roughness (LR in the proposed graph), the increase of ramp slope implies a decrease of energy dissipation. This is mainly due to the fact that reducing ramp slope, and for constant H, the dissipative process is occurring on a larger length resulting in a reduced residual energy.

The previous analysis proved that the relative energy dissipation is a monotonic decreasing function of both S and k/d_{50}. This occurrence was clearly represented by Pagliara and Chiavaccini (2006a) by the following proposed equation:

$$\Delta E_r = A + (1 - A)e^{(B+CS)\frac{k}{H}}$$ (2.3)

In which A, B and C are three coefficients depending on the scale roughness. Namely, for SR conditions $A = 0.15$, $B = -1.0$ and $C = -11.5$; for IR conditions $A = 0.25$, $B = -1.2$ and $C = -12.0$; for LR conditions $A = 0.33$, $B = -1.3$ and $C = -14.5$. In addition, the authors furnished a set of coefficients valid for smooth ramp: $A = 0.02$, $B = -0.9$ and $C = -25.0$. The proposed Eq. (2.3) is valid for S ranging between 0.08 and 0.25, for $k/d_{50} < 42$ and for $k/H < 1.2$.

2.3 ENERGY DISSIPATION ON BASE BLOCK RAMPS IN DIFFERENT SUBMERGENCE CONDITIONS

The previous paragraph focused on the energy dissipation occurring on the ramp itself, in the case of a free hydraulic jump entirely located in the stilling basin. In this paragraph, the analysis of the global dissipative process occurring in association with a block ramp is taken into consideration. Namely, the global dissipative process is the energy dissipation between the ramp entrance and the section downstream of the hydraulic jump (from section 0 to section 2). This phenomenon was deeply analysed by Pagliara et al. (2008). They conducted experimental tests over a wide range of hydraulic and geometric parameters. In particular, they analysed the energy dissipation between section 0 and section 2 for different hydraulic jump submergence ratio L/L_T, where L_T is the horizontal length of the ramp and L is the distance of the beginning of the hydraulic jump from the ramp toe (see Figure 2.4a–d). Experimental tests were conducted varying ramp slope S from 0.125 to 0.25, $k/H < 1.2$ and $k/d_{50} < 42$. In addition, four different ramp bed materials and two different stilling basin materials were used. All the materials employed in the laboratory model were uniform. In order to investigate also the effect of movable bed in the dissipative process, the authors adopted the following experimental methodology. Maintaining both the ramp slope and discharge constant, they evaluated the energy dissipation between section 0 and 2, in the case in which free hydraulic jump ($L/L_T = 0$) and two

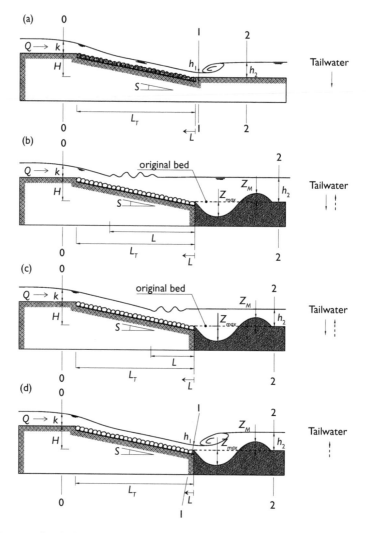

Figure 2.4 Diagram sketch illustrating (a) free jump and fixed downstream stilling basin bed; submerged ramp condition with (b) $L/L_T = 0.66$, (c) $L/L_T = 0.33$; (d) $L/L_T = 0$ (free jump in mobile bed).

submerged hydraulic jump configurations ($L/L_T = 0.33$ and 0.66) was occurring, both in the case of presence and absence of the scour hole downstream of the ramp toe. This was achieved by firstly simulating the submerged condition $L/L_T = 0.66$ and then, by lowering the tailwater, the other two hydraulic jump configurations were obtained ($L/L_T = 0.33$ and $L/L_T = 0$). In this "decreasing tailwater" phase no scour took place. Once a free hydraulic jump was established, the scour process also occurred and a scour hole formed downstream of the ramp toe. Successively, by increasing again the tailwater level, the previous hydraulic jump submerged configurations ($L/L_T = 0.33$ and 0.66) were reproduced, in order to check the effect of the scour hole formation on the dissipative process (all other variables being kept constant). The experimental methodology is sketched in the following Figure 2.4a–d.

For each configuration, the authors estimated both total upstream energy E_0 and total energy at section 2, $E_2 = h_2 + V_2^2/(2g)$. The authors showed that the functional relationship governing the global dissipative phenomenon should be expressed as follows:

$$\Delta E_2 = f\left(\frac{k}{d_{50}}, \frac{k}{H}, S, \frac{L}{L_T}\right) \qquad (2.4)$$

where $\Delta E_2 = (E_0 - E_2)/E_0$ and is the relative energy dissipation between section 0 and section 2. Their analysis aimed to highlight the effect of each non-dimensional group present in the functional relationship Eq. (2.4) on the dependent variable ΔE_2. In particular, they compared the residual energy E_2 associated with the same hydraulic jump configuration and for the same hydraulic conditions for both the cases of absence and presence of a scour hole downstream of the ramp toe. That is, the authors compared the energy dissipation $(E_0 - E_2)$ for the same configuration of the hydraulic jump $(L/L_T = 0.33$ and $0.66)$ in the corresponding decreasing and increasing tailwater phases. They showed that the effect of scour presence on the dissipative process is practically negligible, i.e. $\Delta E_{2\,dw} \approx \Delta E_{2\,up}$, where $\Delta E_{2\,dw} = \Delta E_2$, in the case in which the tailwater level was decreased and in the absence of a scour hole in the stilling basin, and $\Delta E_{2\,up} = \Delta E_2$, in the case in which the tailwater level was increased and in the presence of a scour hole in the stilling basin. In addition, the authors proved other two important issues for submerged hydraulic jump configurations $(L/L_T = 0.33$ and $0.66)$: 1) $\Delta E_{2\,mobile} \approx \Delta E_{2\,smooth}$, i.e., the dissipative process is essentially the same, for identical conditions and hydraulic jump configurations, in the cases in which the downstream stilling basin is either smooth and rigid or movable; 2) in the tested ranges of ramp slope $(0.125 < S < 0.25)$, the effect of S on ΔE_2 is negligible, for the same relative roughness. These two observations assume a fundamental importance, as they indicate that the functional relationship expressed by Eq. (2.4) can be re-written as follows:

$$\Delta E_2 = f\left(\frac{k}{d_{50}}, \frac{k}{H}, \frac{L}{L_T}\right) \qquad (2.5)$$

In particular, the previous observation (1) allows to state that the dissipative process is mainly occurring on the ramp itself, and the scour presence is negligible in terms of dissipation for submerged jump conditions. This occurrence can be easily inferred from the following Figure 2.5, in which, for a fixed slope, the relative global energy dissipation occurring either in presence of a mobile or a smooth fixed bed is compared for different relative roughness conditions.

All the previous deductions are relative to submerged jump configurations, but the authors also showed that if a free hydraulic jump is occurring downstream of the ramp in a mobile bed, the global energy dissipation depends marginally on both ramp slope and relative roughness of the ramp. In this case, it means that the most part of the dissipative process is associated with the occurrence of the hydraulic jump in the stilling basin, i.e., it is depending only on k/H, as proved by the following figure

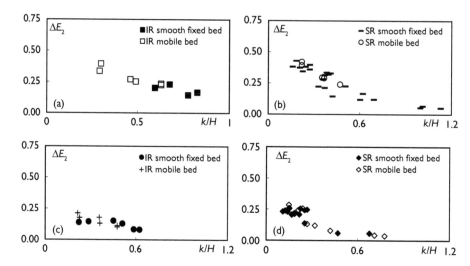

Figure 2.5 $(E_0 - E_2)/E_0$ vs k/H for (a) $S = 0.125$ and $L/L_T = 0.33$, (b) $S = 0.166$ and $L/L_T = 0.33$, (c) $S = 0.166$ and $L/L_T = 0.66$, (d) $S = 0.25$ and $L/L_T = 0.66$.

in which experimental data of all the tested relative roughness configurations and slopes are reported.

In conclusion, Pagliara et al. (2008) stated that the global energy dissipation can be satisfactorily expressed and estimated using the following equation:

$$\Delta E_2 = A + (1 - A)e^{(B)k/H} \tag{2.6}$$

In which A and B are parameters depending on scale roughness (SR, IR, LR) and hydraulic jump position (L/L_T). Note that for the free hydraulic jump $L/L_T = 0$. A and B assume the following expressions according to different scale roughness:

$$A = 0.239e^{-2.323\left(\frac{L}{L_T}\right)} \tag{2.7a}$$

$$B = -\left(10.7\frac{L}{L_T} + 1.729\right) \tag{2.7b}$$

valid for SR roughness condition,

$$A = 0.249e^{-1.618\left(\frac{L}{L_T}\right)} \tag{2.8a}$$

$$B = -\left(9.95\frac{L}{L_T} + 1.863\right) \tag{2.8b}$$

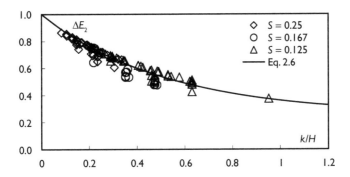

Figure 2.6 $(E_0 - E_2)/E_0$ vs k/H for free jump and mobile stilling basin.

valid for IR roughness condition and

$$A = 0.256e^{-1.245\left(\frac{L}{L_T}\right)} \qquad (2.9a)$$

$$B = -\left(8.475\frac{L}{L_T} + 1.931\right) \qquad (2.9b)$$

valid for LR roughness condition. Also, for the global dissipative process, the effect of the scale roughness is significant. In fact, for all the tested submerged configurations, higher dissipation occurs for LR and it decreases for IR and SR, for all other parameters held constant. The previous equations are valid in the following ranges of parameters: $0 < L/L_T < 0.7$, $0.1 < k/H < 1.2$, $0.125 < S < 0.25$ and roughness condition SR, IR and LR. For the case of a free hydraulic jump occurring in a mobile bed, the authors found that, by assuming $A = 0.25$ and $B = -1.9$, data are well predicted by Eq. (2.6). This agreement is shown in Figure 2.6, for all the tested roughness and geometric configurations.

2.4 HYDRAULIC JUMP CHARACTERISTICS DOWNSTREAM OF BLOCK RAMPS

According to the findings reported in the previous paragraph, great energy dissipation is associated with the occurrence of the hydraulic jump. Thus, fundamental importance is attached to which type of hydraulic jump is occurring on the ramp or downstream of it and how the flow pattern in the stilling basin varies because of the geometric configuration. The study of the hydraulic jump was conducted by several authors. Namely, the first study of the hydraulic jump characteristics downstream of a block ramp was proposed by Pagliara (2007) for uniform stilling basin material and successively extended by Palermo et al. (2008). In the case of a prismatic stilling basin having the same width of the ramp, two different hydraulic jump types can occur, in general. Pagliara (2007) and Palermo et al. (2008) termed them as S_{MB} and

F_{MB} jump types. S_{MB} jump type (Submerged jump in Mobile Bed) is characterized by a counter-clockwise flow circulation, as shown in the following Figure 2.7. Namely, the flow tends to submerge the ramp toe and sediment transport direction is only towards downstream. The second jump type is termed F_{MB} jump type (Free jump in Mobile Bed) and is characterized by a clockwise flow circulation (Figure 2.8). An oscillating behaviour takes place, up to when the equilibrium scour configuration is reached, i.e., the hydraulic jump configuration changes during the erosive process, passing from S_{MB} to F_{MB} typology up to when it becomes stable. The sediment transport direction is characterized by a different behaviour respect to the previous cited typology. Sediment are transported both downstream and upstream of the scour hole, such that the scour hole appears flatter and longer than that occurring in the presence of a S_{MB} jump type.

Furthermore, Palermo et al. (2008) proposed a parametric classification in terms of densimetric Froude number F_{d90} and ramp slope S, for different material non-uniformity coefficient values σ, by which it is possible to predict which hydraulic jump typology will occur (Figure 2.9).

In particular, in this graph, according to different σ values, four fields are distinguished in terms of densimetric Froude number and ramp slope. Namely,

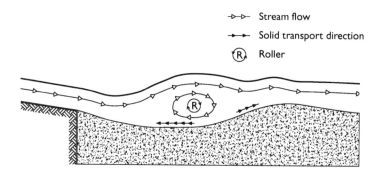

Figure 2.7 Diagram sketch of S_{MB} jump type.

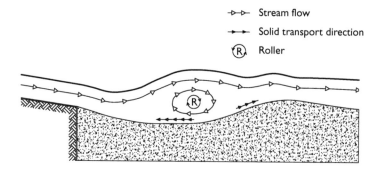

Figure 2.8 Diagram sketch of F_{MB} jump type.

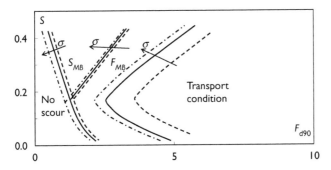

Figure 2.9 Existence fields of hydraulic jumps occurring downstream of a block ramp for $1 < \sigma < 2.8$.

the "*No scour*" region is characterized by the absence of any erosive process. It means that the energy of the flow is not enough to remove sediment from the stilling basin. As shown in the previous Figure 2.9, it occurs for low densimetric Froude numbers. In addition, both the effect of stilling basin material non-uniformity and ramp slope are evident. The scour process takes place for lower F_{d90} if the ramp slope increases. The same effect can be detected by increasing material non-uniformity. The "*Transport condition*" occurs when there is no ridge formation downstream of the scour hole, thus the scour process is not confined by the ridge and the scour hole proceeds downstream becoming longer and flatter. This condition is reached for higher densimetric Froude numbers. The effect of the sediment non-uniformity on the delimiting lines of the various existence fields is to shift the curves towards lower F_{d90}, i.e., for constant ramp slope, all the region shift towards lower F_{d90} if σ increases. Palermo et al. (2008) tested several material whose σ ranged between 1.2 (uniform) and 2.77. Namely, they tested three different materials having the same mean diameter d_{50} and different σ: material 1 with $\sigma_1 = 1.2$; material 2 with $\sigma_2 = 1.79$; and material 3 with $\sigma_3 = 2.77$. They concluded that the "*no scour*" region can be reasonably represented by the following parametric equation:

$$S_0 = -0.35F_{d90} + 0.02\sigma^2 - 0.15\sigma + 0.8 \qquad (2.10)$$

which is valid for the following range of parameters: 1V:12H < S < 1V:4H and $1 < \sigma < 2.8$, whereas the "*transport condition*" existence field can be limited by two straight lines whose equations are:

$$S = -0.1F_{d90} + 0.05\sigma^2 - 0.29\sigma + 0.8 \qquad (2.11a)$$

for 1V:12H < S < 1V:6H and $1 < \sigma < 2.8$ and

$$S = 0.09F_{d90} - 0.05\sigma^2 + 0.28\sigma - 0.41 \qquad (2.11b)$$

for 1V:6H < S < 1V:4H and $1 < \sigma < 2.8$. Finally, F_{MB} region is the field in which the F_{MB} hydraulic jump type is occurring, whereas S_{MB} region is characterized by the formation of a S_{MB} hydraulic jump type. The existence fields of these two hydraulic jump typologies depend on F_{d90} and S and vary according to the material tested (σ). If the downstream stilling basin has not the same width of the ramp, but an abrupt symmetrical enlargement is present downstream of it, the flow pattern (and consequently the hydraulic jump structure) becomes more complex and three-dimensional. That is, according to the different ratios B/b (where B is the stilling basin width and b is the ramp width) and to the hydraulic conditions and ramp configuration, several 3D hydraulic jumps can take place. In the following Figure 2.10 a sketch of an enlarged stilling basin downstream of a block ramp, including the transversal cross-section, is reported. The following sketch illustrates the experimental apparatus adopted by both Pagliara et al. (2009a) and Pagliara et al. (2009b). Namely, they tested different enlargement ratios (B/b = 1, 1.8, 2.8), in the presence of both uniform and non-uniform sediment in the stilling basin and for S varying between 1V:8H and 1V:4H, concluding that four different hydraulic jump types can occur in the specified configurations, according to the different hydraulic conditions: *Repelled Symmetric jump* (R-S); *Repelled Oscillatory jump* (R-O); *Toe Symmetric jump* (T-S); and *Toe Oscillatory jump* (T-O). Both the flow patterns and corresponding scour morphologies are sketched in Figure 2.10. The existence fields of the cited hydraulic jump types can be found in details in Pagliara et al. (2009a). In the following, just a brief description of the various jump characteristics will be furnished.

Repelled Symmetric jump (R-S, Figure 2.11a) is a 3D undular hydraulic jump characterized by superficial shock waves and lateral flow re-circulation zones located symmetrically. The resulting scour morphology is strongly 3D. In the case of *Repelled Oscillatory jump* (R-O, Figure 2.11b), the flow structure is characterized by periodical lateral oscillations which can eventually reach both stilling basins side walls, resulting in a less 3D scour hole morphology with respect to the previous case. *Toe Symmetric*

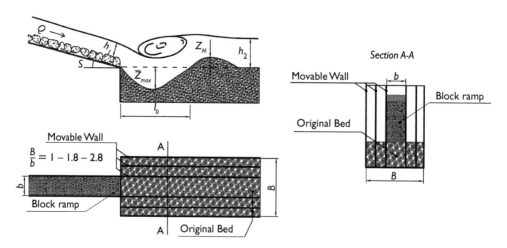

Figure 2.10 Diagram sketch of the experimental apparatus for symmetrically enlarged stilling basin.

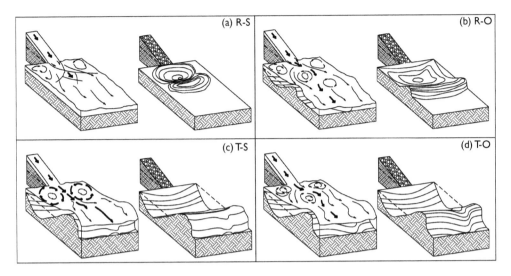

Figure 2.11 Sketches of flow patterns and equilibrium scour morphology for 3D hydraulic jump types.

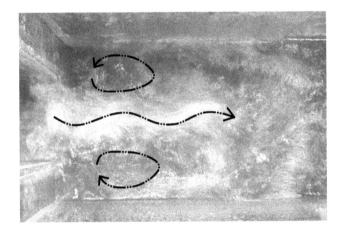

Figure 2.12 Picture showing a T-S jump type.

jump (T-S, Figure 2.11c) shows two prominent symmetric lateral re-circulation regions and occurs just downstream of the ramp toe, because of the high downstream tailwater level. The resulting equilibrium scour morphology is characterized by an almost 2D shape and a prominent arched ridge is present. *Toe Oscillatory jump* (T-O) is essentially similar to T-S jump type, as the scour morphology results to be almost 2D, but the flow structure is characterized by periodic oscillations towards both stilling basin side walls. It is evident that (i) different types of jump are both the cause and consequence of the equilibrium scour morphology and (ii) the dissipative process strongly depends on the flow characteristics in the stilling basin. A T-S jump is illustrated in Figure 2.12.

2.5   ENERGY DISSIPATION IN PRESENCE OF SYMMETRICALLY ENLARGED STILLING BASIN

The analysis of energy dissipation in the presence of a symmetrically enlarged basin was conducted by Pagliara et al. (2009b), for both uniform and non-uniform stilling basin materials and for block ramp slopes ranging between 1V:8H and 1V:4H. Based on the findings of Pagliara et al. (2009a) and Pagliara et al. (2008), the authors proved that the global energy dissipation occurring both on the ramp and in the stilling basin (between ramp entrance and the section downstream of the hydraulic jump, see Figs. 2.1 and 2.3) depends on the following non-dimensional parameters:

$$\Delta E_2 = f\left(\frac{k}{H}, \frac{B}{b}, \frac{h_2}{h_1}\right) \tag{2.12}$$

In particular, Pagliara et al. (2009b) analysed the dissipative process in the case in which the hydraulic jump was entirely located in the stilling basin without submerging the ramp and for three different enlargement ratios ($B/b = 1, 1.8, 2.8$). The study of energy dissipation efficiency had been previously conducted by several authors in the presence of an enlarged stilling basin, for the case of a horizontal smooth bed. However, none of these studies (among others Rajaratnam & Subramanya, 1968, Herbrand, 1973, Smith, 1989, Hager, 1985, and Bremen & Hager, 1993) analysed the dissipative process downstream of a block ramp with an enlarged movable stilling basin. Nevertheless, it is useful and interesting to synthetize the results which were deduced by the cited authors. In particular, Bremen and Hager (1993) proved that the hydraulic jump typology occurring in the stilling basin strongly modifies and characterizes the dissipative process. Namely, they stated that the dissipative process depends mainly on the approaching (entering) Froude number $F_1 = U_1/(gh_1)^{0.5}$ and the enlargement ratio (B/b). Even if it was noted that the dissipation rate is generally higher in an enlarged stilling basin (between stilling basin entering section and hydraulic jump end section), Bremen and Hager (1993) proved that a T-jump exhibits lower efficiency than the classical hydraulic jump. The presence of a mobile bed contributes to strongly modify the flow structure and consequently it deeply affects the dissipative process. This occurrence is mostly due to the formation of a scour hole that, according to the different hydraulic conditions and geometric configurations, can be strongly 3D or almost 2D. In particular, a fundamental role is played by the parameter h_2/h_1, as it contributes to modify the non-dimensional transversal scour profile and consequently the hydraulic jump characteristics inside the stilling basin (Pagliara et al., 2009a). Vice versa, the effect of h_2/h_1 on the dissipative process for $B/b = 1$ is quite limited due to the fact that the flow structures does not significantly vary, as proved also by both non-dimensional longitudinal and transversal profiles similitudes. Pagliara et al. (2009b) compared the relative global non dimensional energy dissipation $\Delta E_2 = (E_0 - E_2)/E_0$ (see Figures 2.4 and 2.10) for $1 < B/b < 2.8$ and $1 < \sigma < 2.8$. The results of the comparison are shown in Figure 2.13, along with the plot of Eq. (2.6) for which $A = 0.25$ and $B = -1.9$ derived by Pagliara et al. (2008) and valid for $B/b = 1$. From this figure, it can be easily noted that in the case of $B/b = 1$, the results obtained by Pagliara et al. (2008) are confirmed, meaning that the energy dissipation process is not influenced by

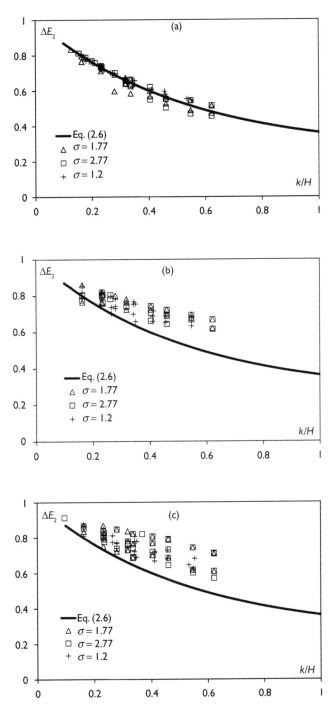

Figure 2.13 $(E_0 - E_2)/E_0$ versus k/H for different σ, S and B/b = (a) 1, (b) 1.8, (c) 2.8.

both material non-uniformity and block ramp slope. Whereas in the case of $B/b > 1$, the data appear more spread for the reasons illustrated above (effect of both B/b and h_2/h_1). In particular, the effect of material non-uniformity is still not prominent, but the energy dissipation process results in being deeply influenced by the parameters B/b and h_2/h_1. That is, Pagliara et al. (2009b) stated that the effect of h_2/h_1 is more prominent increasing B/b. Nevertheless, experimental tests confirmed the findings of Hager (1985), i.e., energy dissipation generally increases for higher expansion ratios. Based on these observations and deductions, Pagliara et al. (2009b) further developed the findings of Pagliara et al. (2008) and extended the validity of Eq. (2.6) to the following ranges of non-dimensional parameters: $1 < \sigma < 2.8$, $0.16 < k/H < 0.62$, h_2/h_1 up to 2.87 and $0.125 < S < 0.25$.

They proposed the following general equation:

$$\Delta E_2 = A + (1-A)e^{(C)k/H} \tag{2.13a}$$

in which

$$A = \frac{\alpha \dfrac{h_2}{h_1} + \beta}{1 - e^{(Ck/H)}} \tag{2.13b}$$

where

$$\alpha = f(B/b, k/H) = -0.084(1 - e^{(-10k/H)})(B/b - 1)^{0.36} \tag{2.13c}$$

$$\beta = f(B/b, k/H) = 0.25(1 - e^{(Ck/H)}) + [0.232(1 - e^{(-10k/H)})(B/b - 1)^{0.36}] \tag{2.13d}$$

and $C = -1.9$. It can be easily noted that for prismatic channels ($B/b = 1$), Eq. (2.13a) coincides with Eq. (2.6), as $A = 0.25$ and $C = -1.9$. It means that it can be considered the general equation to predict global relative energy dissipation for enlargement ratios up to 2.8. All the previous deductions are valid in the case in which $F_1 > 1 > F_2$, where $F_2 = U_2/(gh_2)^{0.5}$ is the Froude number at the end of the hydraulic jump. Furthermore, the energy dissipation process occurring inside the stilling basin (between section 1 and 2, see Figure 2.4) in the case of mobile bed is more prominent than that corresponding to a horizontal fixed bed. In fact, a comparison of experimental data of the variable $[(E_1 - E_2)/E_1]_{meas}$ (energy dissipation in the stilling basin) relative to enlarged channels in the presence of a movable bed with the corresponding predicted values $[(E_1 - E_2)/E_1]_{calc}$ obtained using the formula proposed by Hager (1985), shows that, generally, $[(E_1 - E_2)/E_1]_{meas} > [(E_1 - E_2)/E_1]_{calc}$. It means that for identical hydraulic and geometric configuration of the stilling basin, the presence of a mobile bed contributes to increase energy dissipation. This can be easily explained considering that part of the flow energy is dissipated in the erosive process. Finally, this effect is amplified if B/b increases, because scour depth generally increases for higher expansion ratios, as proved by Pagliara et al. (2009a).

2.6   ENERGY DISSIPATION IN PRESENCE OF ASYMMETRICALLY ENLARGED STILLING BASIN

Energy dissipation process downstream of a block ramp is deeply influenced by stilling basin geometry, as shown in the previous paragraph. In addition, it was shown also that the flow pattern and hydraulic jump type plays an essential role in the dissipative mechanism. Nevertheless, the previous paragraph focused on a specific stilling basin configuration, i.e. symmetrically enlarged. But, in practical application it can happen that both the natural location of the structure or other external conditions do not allow the building of a symmetrically enlarged stilling basin. Thus, it is not rare to find asymmetrically enlarged stilling basins. This last configuration presents some similarities but even fundamental differences with respect to that illustrated above. The following Figure 2.14 shows a sketch illustrating an asymmetrically enlarged stilling basin, along with the indication of the main hydraulic and geometric parameters.

This last configuration was tested by Pagliara and Palermo (2012). The authors analysed the global dissipative process occurring both on a block ramp and in the stilling basin for two different enlargement ratios of the stilling basin. They conducted experiments for $B/b = 2.33$ and 1.67, including some preliminary tests for $B/b = 1$, adopting an experimental apparatus that is schematically reported in the previous figure. In addition, tests were conducted for LR, IR and SR conditions (Pagliara & Chiavaccini, 2006a), for both uniform and non-uniform stilling basin materials and for a single ramp slope equal to $S = 0.125$. Also in this case, it is fundamental to analyse and illustrate the flow behaviour in the stilling basin. In particular, the description of the hydraulic jump types occurring inside it and their respective characteristics can be useful to understand both similarities and differences between symmetrically and asymmetrically enlarged basins. In the tested range of parameters ($2.5 < F_{d90} < 4.5$, $S = 0.125$, $\sigma < 2.8$ and $h_2/h_1 < 2.6$), Pagliara and Palermo (2012), showed that two

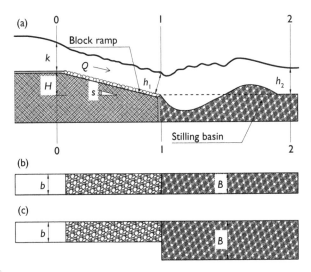

Figure 2.14 Diagram sketches illustrating (a) profile view; plan view for (b) $B/b = 1$ (no expansion) and (c) asymmetric expansion.

different jump types can occur and they termed them *Repelled Asymmetric* (R-A) and *Repelled Asymmetric Oscillating* (R-A-O), respectively. It has to be noted that a symmetrically-enlarged configuration can be considered as the union of two identical, asymmetrically-enlarged configurations, whose un-enlarged channel wall can be considered as the vertical plane of symmetry. A brief description of these two hydraulic jumps characteristics is essential for the following considerations. In Figure 2.15 the two different hydraulic jumps are sketched schematically. Figure 2.16 shows a picture of an R-A-O jump type.

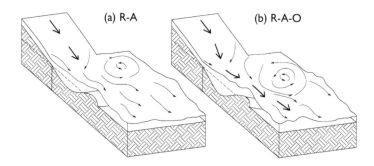

Figure 2.15 Diagram sketch of the (a) R-A and (b) R-A-O hydraulic jump types.

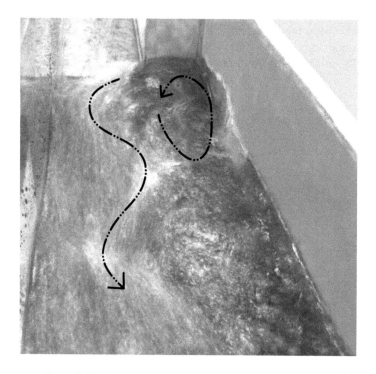

Figure 2.16 Picture illustrating a R-A-O hydraulic jump type.

The R-A jump (Fig. 2.15a) appears similar to the R-S jump occurring in the symmetrically enlarged stilling basin. In fact, both these types of hydraulic jumps show superficial shock waves, they are both generally undular and occur for relatively low F_{d90} for the tested bed ramp slope. A stable lateral flow re-circulation occurs also in the case of the asymmetrically-enlarged basin, resulting in a strong 3D scour hole and ridge whose maximum depth and height, respectively, are located inside the basin itself. Thus, even if the flow circulation and structure is similar to that characterizing the R-S jump, the scour morphology exhibits significant differences. The R-A-O type (Figs. 2.15b and 2.16) shows similarities with the R-O jump as it is characterized by a huge lateral flow recir-culation and, in addition, periodical oscillations are visible on the free surface, rarely reaching lateral walls. The hydraulic jumps occurring downstream of both asymmetrically- and symmetrically-enlarged stilling basins are also characterized by similar existence fields, as shown in the following Figure 2.17a–b, valid for $S = 0.125$ and for asymmetric and symmetric configurations of the stilling basin, respectively. In the case of asymmetrically enlarged basin, such as for the sym-metric case, for $1 < B/b < 1.2$, the flow structure is essentially 2D, as proved also by Nashta and Garde (1988), resulting in a F_{MB} jump type. For $1.2 < B/b < 1.67$, however, the flow structure exhibits intermediate characteristics between 2D and 3D jump types, this region can, thus, be considered a transition field. Increasing the enlargement ratio causes an increase of lateral flow recirculation resulting in a more prominent 3D flow structure (and morphology in the stilling basin). The transition between the two hydraulic jump typologies takes place for $F_{d90} \approx 3.7$ for $B/b = 1.67$ and $F_{d90} \approx 3.9$ for $B/b = 2.33$, as shown in Figure 2.17a. Comparing the existence fields of R-S and R-O and of R-A and R-A-O in Figure 2.17a–b, it is evident that the transition for the asymmetric case occurs for higher densimetric Froude numbers. Based on these observations, Pagliara and Palermo (2012), ana-lysed the global dissipative process also in the case of an asymmetrically-enlarged stilling basin.

They showed that, for $S = 0.125$, energy dissipation ΔE_2 for $B/b > 1$ is higher if compared to that relative to the prismatic case ($B/b = 1$) for identical hydraulic con-ditions and ramp configuration, as shown in Figure 2.18a. The general behaviour appears to be similar to that observed for symmetrically-enlarged stilling basins. Also in this case, no clear trend due to different ramp bed roughness can be detected, and the global dissipative process depends both on the enlargement ratio and relative tail-water h_2/h_1, as shown in Figure 2.18a–b.

Thus, even in this case, the functional relationship governing the dissipative phe-nomenon has the same algebraic form of Eqs. (2.6) and (2.13a). Pagliara and Palermo (2012) derived a general equation having the characteristic to be analytically identical to the proposed Eq. (2.6) for $B/b = 1$. In practice, they furnished a general equation valid for asymmetrically-enlarged stilling basins which coincides with Eq. (2.6) in the case of prismatic channels ($B/b = 1$). The authors proposed the following equation (2.14a), valid for $B/b < 2.33$ and $h_2/h_1 < 2.6$:

$$\Delta E_2 = A + (1 - A)e^{B\left[(k/H)f_1\left(B/b, h_2/h_1\right)\right]} \tag{2.14a}$$

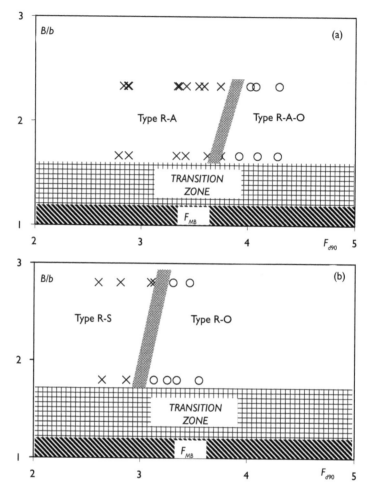

Figure 2.17 Hydraulic jump types classification for $S = 0.125$ and (a) asymmetrically and (b) symmetrically enlarged stilling basin.

where

$$f_1(B/b, h_2/h_1) = (B/b)^{\left[(0.23(B/b)-0.04)(h_2/h_1)+(0.015(B/b)-1.59)\right]}$$ (2.14b)

with $A = 0.25$ and $B = -1.9$. Note that

$$f_1(B/b, h_2/h_1) = 1$$ (2.14c)

for $B/b = 1$. It can be concluded that, in the case of an asymmetrically-enlarged channel, the totality of data can be predicted by Eq. (2.14a), valid in the mentioned ranges of parameters, and the dissipative process is similar to that occurring in the presence of a symmetrically-enlarged channel.

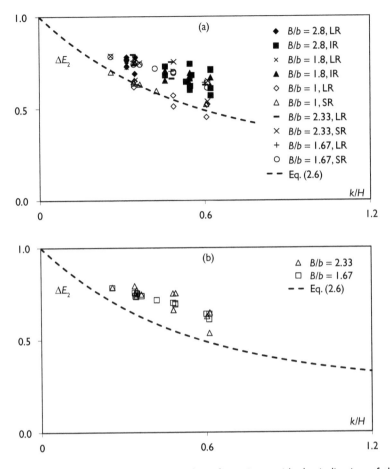

Figure 2.18 (a) ΔE_2 versus k/H for all the tested configurations, with the indication of the relative roughness scale and enlargement ratios; (b) ΔE_2 versus k/H for asymmetrically enlarged configuration ($B/b = 1.67$ and $B/b = 2.33$).

2.7 ENERGY DISSIPATION IN PRESENCE OF PROTECTION MEASURES IN A PRISMATIC STILLING BASIN

The erosive process downstream of a block ramp can be controlled by inserting protection measures in the stilling basin. In particular, rock made sills are efficient countermeasures which can sensibly reduce scour hole geometry if correctly located in the stilling basin. Pagliara and Palermo (2008b) analysed the scour process in the presence of rock made sills located in different spatial positions in a prismatic stilling basin ($B/b = 1$) and, successively, Pagliara and Palermo (2010b) analysed the global dissipative process in the presence of protection measures downstream of the ramp. In particular, their analysis was conducted for both uniform and non-uniform materials constituting the stilling basin and for block ramp slope varying between $0.083 < S < 0.25$. Namely, they adopted protection sills made of blocks whose mean diameter was $D_{50} = 0.046$ m and were located in two vertical and four longitudinal positions, respectively, as shown in the following Figure 2.19a–b.

Pagliara and Palermo (2008b) and Pagliara and Palermo (2010b) conducted a series of experimental tests (reference tests) without any protection structure (see Fig. 2.19a) for various hydraulic conditions and geometric configuration of the ramp. Then, they located the protection structures at $Z_{op} = z_{op}/z_m = 0$ and 0.5 and $\lambda = x_s/l_0 = 0.25, 0.5, 0.75, 1.00$, where Z_{op} is the non-dimensional vertical position and λ the non-dimensional longitudinal position of the structure, with z_{op} as the vertical position of the upper sill corner point measured from the original bed level and x_s is the longitudinal distance of the protection structure from the ramp toe. z_m and l_0 are the maximum average scour depth and scour length, respectively, of the corresponding base (reference) tests (i.e., same hydraulic conditions and configurations). z_{ms} is the maximum scour depth in the presence of a protection structure. Furthermore, tests were conducted for intermediate scale roughness conditions, according to the definition of Pagliara and Chiavaccini (2006a). It was shown experimentally that the global

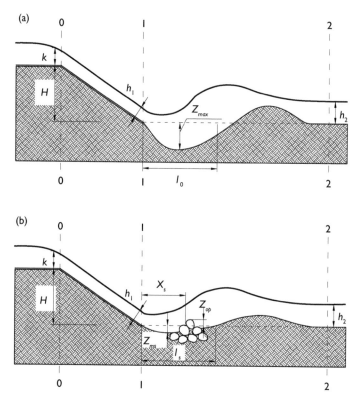

Figure 2.19 Definition sketch of a ramp (a) in absence (*reference tests*) and (b) in presence of protection measures in the stilling basin.

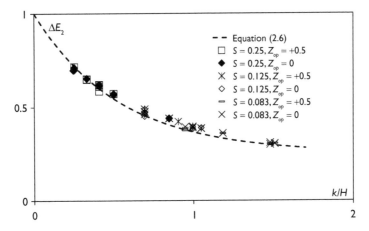

Figure 2.20 ΔE_2 vs *k/H* for all the tested conditions.

relative energy dissipation is not significantly dependent on protection structure presence, as shown in the following Figure 2.20, in which measured data of the variable ΔE_2 are plotted versus *k/H*, along with Eq. (2.6) proposed by Pagliara et al. (2008). In particular, Figure 2.20 reports data relative to different ramp slope tested and longitudinal positions of the protection structure, with the indication of the vertical position and the plot of Eq. (2.6). It can be easily understood that the effect of the protection measure presence slightly affects energy dissipation, i.e., just a slight increase of the global energy dissipation is detectable. But, for practical purposes, it can be considered completely negligible. These findings imply that, in the tested range of parameters, which can be found in Pagliara and Palermo (2010b), Eq. (2.6) can satisfactorily predict the totality of data, including those relative to protected basins. It is obvious that the coefficients *A* and *B* appearing in Eq. (2.6) have to be assumed equal to 0.25 and −1.9, respectively.

2.8   ENERGY DISSIPATION IN THE PRESENCE OF BOULDERS

The dissipative process on a block ramp is deeply influenced by the roughness condition of the ramp itself, as shown in the previous paragraphs. It is clear that a modification of ramp surface implies substantial variation in the dissipative mechanism. However, it is not rare to find block ramps on which boulders are located in order to increase bed stability. Thus, this specific configuration requires a proper attention in order to estimate and evaluate the dissipative process. In particular, boulders can be of different sizes and located in different arrangements on a block ramp. One of the first attempts to analyse the dissipative process on reinforced block ramp was conducted by Pagliara and Chiavaccini (2006b). In particular, they extended the findings of Pagliara and Chiavaccini (2006a), valid for base ramp configuration, to the case in which boulders were located on the ramp in two different arrangements:

in rows and random. They tested several block ramps configurations, whose slope varied between 0.08 and 0.33. Nevertheless, they varied both the boulders' diameter and their surface roughness, i.e. they adopted hemispheres with a protruding part equal to half the diameter D_B, and their surface was both smooth and rough. Rough boulders were obtained by covering them with a sand layer. Finally, they also varied boulders concentration $\Gamma = (N_B \pi D_B^2)/(4bL)$, where N_B is the boulders' number on the ramp, D_B boulders' mean diameter, b the ramp width and L the ramp length. The following deductions are valid in the range of parameters $0.08 < S < 0.33$, $0 < \Gamma < 0.33$, $1.75 < D_B/d_{50} < 19$, with d_{50} as the mean diameter of the ramp base material. In the following Figure 2.21, a reinforced ramp with rough and smooth boulders arranged in rows is shown.

Preliminary observations indicate that the presence of boulders increases energy dissipation respect to the corresponding base configuration. In addition, rough boulders contribute to further increase energy dissipation if compared to smooth boulder counterparts. Another important parameter, which deeply affects dissipative mechanism, is the boulders' concentration. Based on these observations, the functional relationship expressing the increase of energy dissipation in the presence of boulders can be written as follows:

$$\frac{\Delta E_{rB}}{\Delta E_r} = f\left(\frac{d_{50}}{k}, \frac{H}{k}, S, \mathrm{Re}, \frac{D_B}{d_{50}}, F, \varepsilon, disp, \Gamma \right) \tag{2.15}$$

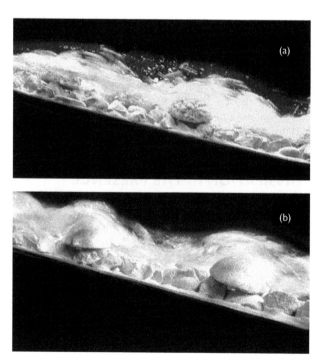

Figure 2.21 Pictures of a reinforced ramp with boulders in rows: (a) rough boulders and (b) smooth boulders.

where $\Delta E_{rB} = (E_0 - E_1)/E_0$ is the relative energy dissipation on a block ramp with protruding boulders (see Figures 2.1 and 2.21), ΔE_r is the relative energy dissipation on the corresponding base block ramp, which can be evaluated using Eq. (2.3), *disp* is the boulders' arrangement (in rows or random), ε boulders' roughness and F the Froude number. Other parameters have been already defined above. Pagliara and Chiavaccini (2006b) proved that the previous Eq. (2.15) can be reduced to the following Eq. (2.16):

$$\frac{\Delta E_{rB}}{\Delta E_r} = f\left(\varepsilon, disp, \Gamma\right) \qquad (2.16)$$

Namely, they found that Eq. (2.15) can be rearranged as follows:

$$\frac{\Delta E_{rB}}{\Delta E_r} = 1 + \frac{\Gamma}{E + F\Gamma} \qquad (2.17)$$

where E and F are coefficients depending on both boulders' arrangement and roughness. For row disposition and rounded boulders $E = 0.55$ and $F = 10.5$; for random disposition and rounded boulders $E = 0.6$ and $F = 13.3$; for random disposition and crushed boulders $E = 0.55$ and $F = 9.1$; for row disposition and crushed boulders $E = 0.4$ and $F = 7.7$. In the following Figure 2.22, plots of Eq. (2.17) are reported. It can be easily observed that random arrangement of boulders causes less energy dissipation with respect to the arrangement in rows due to local acceleration (downstream of each row) and deceleration of the flow, resulting in a localized dissipative mechanism which is more efficient than that occurring with boulders in random arrangement. Finally, it can be observed that Eq. (2.17) tends to 1 if Γ tends to 0 (base configuration). This means that for $\Gamma = 0$, Eq. (2.17) coincides with Eq. (2.3). Based on

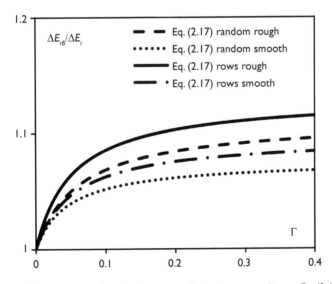

Figure 2.22 Increase in the relative energy dissipation according to Eq. (2.16).

the deductions of Pagliara and Chiavaccini (2006b), other studies were developed by several authors. In particular, Ahmad et al. (2009) analysed the dissipative process occurring on a reinforced block ramp using boulders in staggered arrangement.

These authors tested one ramp slope ($S = 0.25$), whose base material mean diameter was $d_{50} = 0.02$ m, and three different boulders diameters ($D_B = 0.055, 0.065, 0.1$ m). In addition, they conducted experiments varying both the longitudinal and transversal distance of the boulders. For experiments performed with boulders having $D_B = 0.055$ m, three different non-dimensional longitudinal clear spacings and one non-dimensional transversal clear spacing were adopted ($S_x/D_B = 4, 3, 1.5$ and $S_y/D_B = 1.23$); for boulders whose $D_B = 0.065$ m they conducted experiments for $S_x/D_B = 3, 2, 1.5, 1$ and $S_y/D_B = 0.81$; and for boulders whose $D_B = 0.10$ they adopted $S_x/D_B = 2, 1.5, 1$ and $S_y/D_B = 1$. S_x and S_y are the longitudinal and transversal clear spacings between the boulders. Ahmad et al. (2009) observed that energy dissipation increases if S_x/D_B increases. Furthermore, they found that, for constant spacing between the boulders, the dissipative mechanism is affected by the boulders diameter D_B, i.e., it increases with D_B. This last finding led the authors to state that the dissipative mechanism in the presence of a staggered configuration of the boulders is quite different from that observed for both arrangements in rows and random, as it is depending also on D_B. The resulting functional relationship in the specified range of parameters, including $0.074 < \Gamma < 0.21$ and $0.506 < D_B/k < 2.307$, is:

$$\frac{\Delta E_{rB}}{\Delta E_r} = f\left(\frac{D_B}{k}, \Gamma\right) \tag{2.18}$$

Ahmad et al. (2009) stated that, to estimate the relative energy dissipation for staggered configuration, an equation formally identical to Eq. (2.17) can be used. They showed that E can be assumed equal to 0.6 and $F = 7.9(D_B/k)^{-0.9}$, thus the final relationship becomes:

$$\Delta E_{rB} = \left[A + (1-A)e^{(B+CS)(k/H)}\right] \cdot \left(1 + \frac{\Gamma}{0.6 + 7.9\left(D_B/k\right)^{-0.9}\Gamma}\right) \tag{2.19}$$

Note that A, B, C are the same as in Pagliara and Chiavaccini (2006a). Another study dealing with reinforced block ramps was that proposed by Oertel and Schlenkhoff (2012). The authors studied crossbar block ramps, which are peculiar structures in which boulders are arranged in transversal rows, forming several basins on the ramp itself. In addition, the rows are made of large boulders inside which some lower stones are located, in order to allow fish migration and overfall for small discharges. Oertel and Schlenkhoff (2012) simulated a crossbar ramp in a laboratory model with slope is $S = 1/30$. They used boulders whose height was $h_b = 6$ cm (to simulate large boulders) and $h_{b,S} = 2$ cm for small boulders. The total opening width created by small boulders inside the row was $b_w = 0.22$ m and the ramp width was $b = 0.8$ m. The base material used for the ramp had a mean diameter $d_{50} = 2$ mm. In the model, seven rows (crossbars), whose relative distance was 0.33 m, were simulated. Furthermore, the deductions obtained by using the physical model were adopted to calibrate a numerical model, by which the investigated range of configurations and parameters has been extended. In particular, the authors analysed the energy dissipation process occurring on this type of structures. In the tested range of parameters (including both numerical and physical experiments), the authors proved that the effect of the boulders' roughness and bed material can be considered negligible. This is mainly due to the fact that the effects of these two parameters were not analysed in details, as the main part of the dissipation occurs because of the presence of the boulders. Thus, for a crossbar ramp, in which the main flow resistance is due to the crossbars themselves, the authors concluded that Eq. (2.17) proposed by Pagliara and Chiavaccini (2006b) can be simplified as follows:

$$\Delta E_{rB} = A + (1 - A)e^{(B+CS)\frac{k}{H}}$$
(2.20)

where $A = 0.17 - 0.0017/S$, $B = -0.7 + 0.0073/S$ and $C = -4.9 - 0.26/S$. However, the previous relationship is valid only for the configurations tested both physically and numerically and it is subjected to the following limitation $1/30 < S < 1/10$ regarding the ramp slope.

2.9 DISSIPATIVE PROCESS IN THE PRESENCE OF SEDIMENT TRANSPORT ON A BLOCK RAMP

Normally, sediment transport characterizes natural rivers. Sediment transported downstream interacts with structures, contributing to modifying both scour processes and dissipative mechanisms. In particular, in the case of block ramps, filling material varies ramp surface characteristics/morphology. It implies that the relative roughness can be affected by sediment transport and, consequently, energy dissipation process is influenced. Pagliara et al. (2009c) tested the effects of sediment transport on block ramps. They conducted experiments for block ramps whose slope varied in the range $0.125 < S < 0.25$, with three different ramp material. They simulated the filling process using three different materials, which were the same used for the stilling basin. The ratio d_{50}/D_{50} of the mean diameters of the block ramp (D_{50}) and of the filling (stilling basin) materials (d_{50}), respectively, varied between 0.073 and 0.298. In the experimental tests, filling material was supplied at the ramp entrance and it was transported downstream on the ramp by the approaching flow. Pagliara et al. (2009c) adopted the following methodology to perform their experiments. At the beginning of each test, a very low discharge Q_1 was supplied along with filling sediment. In this phase, just the upstream part of the ramp was filled. The discharge was gradually increased, keeping constant the sediment supply. Gradually the ramp became completely filled. Then, the discharge was further increased, resulting in a decrease of the sediment level trapped between ramp blocks, up to when the dynamic equilibrium configuration of the ramp surface was reached. That is, for a certain discharge, termed Q_F (filling discharge), the filling sediment level in the ramp did not vary and the exiting sediment quantity from the ramp toe equalled the supplied sediment quantity at the ramp entrance. If the discharge is further increased, all the supplied material is transported downstream and the filling sediment level in the ramp decreases. Finally, when the discharge is equal to Q_E (empty discharge), only the filling material which is trapped or hidden by ramp blocks remains on the ramp itself and any further increase does not modify the filling sediment level significantly. In the following Figure 2.23a–d, a top view of a filled ramp is shown for different discharges.

Pagliara et al. (2009c) analysed the dissipative process when the ramp is filled, i.e. for $Q > Q_F$, and compared the relative energy dissipation $\Delta E_{r(filled)} = (E_0 - E_1)/E_0$ with the corresponding one obtained in the absence of sediment filling $\Delta E_{r(empty)}$ (which can be computed also using Eq. 2.3). They evaluated the variable $\varepsilon = \Delta E_{r(filled)} - \Delta E_{r(empty)}$ and proved that differences are negligible if $Q > Q_F$, as shown by the following graph.

This occurrence is mainly due to the fact that the base relative roughness scale does not substantially vary if $Q > Q_F$, i.e., the scale roughness condition is practically the same. It can be observed that in the tested range of parameters $\varepsilon < 4\%$, thus it can be considered negligible for practical application and Eq. (2.3), valid for a clean ramp, can be applied also in this case.

Figure 2.23 (a)–(d) Top view of the evolution of the filling of ramp central sectors.

Figure 2.24 ε versus k/H for all the tests with $Q > Q_F$.

2.10 SUMMARY

This chapter reports a critical summary of the main findings present in scientific literature up to date regarding the dissipative processes occurring in association with block ramps. In particular, the dissipative process was discussed and analysed for several ramp and stilling basin configurations. Firstly, the dissipative mechanism in the presence of a base ramp was discussed. It was shown that, in the case of a hydraulic jump entirely located in the stilling basin, the mechanism mainly depends on several parameters: the relative roughness, the relative ramp height and the ramp slope. If the ramp is submerged and the hydraulic jump is totally or partially occurring on it, the global dissipation is influenced mainly by the relative scale roughness and the hydraulic jump position. It was also shown that the global dissipative process (including the amount of energy dissipated by the hydraulic jump) is strongly dependent on the hydraulic jump type and location. Thus, the various hydraulic jump typologies were introduced and illustrated, both in the case of prismatic and enlarged stilling basins (asymmetrically and symmetrically). For both symmetrically- and asymmetrically-enlarged stilling basins, the dissipative mechanism was discussed and compared to that occurring in the corresponding prismatic configuration. In addition, the energy dissipation occurring in association with reinforced block ramps was extensively analysed. In particular, the effect of the presence of boulders located in different arrangements was discussed. It was proven that, whatever the boulders' arrangement, the energy dissipation increases with respect to the corresponding base configuration for the reported ranges of boulders' concentration. Finally, the energy dissipation occurring on a filled ramp was briefly described, concluding that the variation respect to the reference ramp configuration is generally limited. This chapter aims to furnish a synthesis of the main research achievements on the topic, but the complexity and variety of the configurations and hydraulic conditions which can occur in practical applications requires further investigations and researches. In particular, most of the proposed relationships are empirical, thus they are valid in the ranges of parameters and for the geometric configurations which were tested by various authors. Any extrapolation or extension of the validity ranges of the proposed formulae should be carefully verified and tested. Nevertheless, the authors hope that this chapter could be a stimulus for researchers to conduct further studies which can help to better understand a so complex phenomenon.

REFERENCES

Ahmad, Z., Petappa, N.M. & Westrich, B. (2009). Energy dissipation on block ramps with straggled boulders. *Journal of Hydraulic Engineering*, 135 (6), 522–526.
Bhaturst, J.C. (1978). Flow resistance of large-scale roughness. *Journal of Hydraulic Division*, 104 (12), 1587–1603.
Bhaturst, J.C., Li, R.M. & Simons, D.B. (1981). Resistance equation for large scale roughness. *Journal of Hydraulic Division*, 107 (12), 1593–1613.
Bremen, R. & Hager, W.H. (1993). T-jump in abrupt expanding channel. *Journal of Hydraulic Research*, 31 (1), 61–78.
Chamani, M.R. & Rajaratnam, N. (1999a). Characteristic of skimming flow over stepped spillways. *Journal of Hydraulic Engineering*, 125 (4), 361–368.

Chamani, M.R. & Rajaratnam, N. (1999b). Onset of skimming flow over stepped spillways. *Journal of Hydraulic Engineering*, 125 (9), 969–971.

Chanson, H. (1994). Hydraulic design of stepped cascades, channels, weirs and spillways. Oxford, Pergamon.

Christodoulou, G.C. (1993). Energy dissipation on stepped spillways. *Journal of Hydraulic Engineering*, 119 (5), 644–650.

Diez-Cascon, J., Blanco, J.L., Revilla, J. & Garcia, R. (1991). Studies on the hydraulic behavior of stepped spillways. *International Water Power & Dam Construction*, 43 (9), 22–26.

Ferro, V. (1999). Evaluating friction factor for gravel bed channel with high boulder concentration. *Journal of Hydraulic Engineering*, 125 (7), 771–778.

Herbrand, K. (1973). The spatial hydraulic jump. *Journal of Hydraulic Research*, 11 (3), 205–218.

Hager, W.H. (1985). Hydraulic jump in non-prismatic rectangular channels. *Journal of Hydraulic Research*, 23 (1), 21–35.

Lawrence, D.S.L. (1997). Macroscale surface roughness and frictional resistance in overland flow. *Earth Surface Processes and Landforms*, 22, 365–382.

Nashta, C.F. & Garde, R.J. (1988). Subcritical flow in rigid-bed open channel expansion. *Journal of Hydraulic Research*, 26 (1), 49–65.

Oertel, M. & Schlenkhoff, A. (2012). Crossbar block ramps: flow regimes energy dissipation, friction factors, and drag forces. *Journal of Hydraulic Engineering*, 138 (5), 440–448.

Pagliara, S. & Chiavaccini, P. (2006a). Energy dissipation on block ramps. *Journal of Hydraulic Engineering*, 132 (1), 41–48.

Pagliara, S. & Chiavaccini, P. (2006b). Energy dissipation on reinforced block ramps. *Journal of Irrigation and Drainage Engineering*, 132 (3), 293–297.

Pagliara, S. (2007). Influence of sediment gradation on scour downstream of block ramps. *Journal of Hydraulic Engineering*, 133 (11), 1241–1248.

Pagliara, S. & Palermo, M. (2008a). Scour control downstream of block ramps. *Journal of Hydraulic Engineering*, 134 (9), 1376–1382.

Pagliara, S. & Palermo, M. (2008b). Scour control and surface sediment distribution downstream of block ramps. *Journal of Hydraulic Research*, 46 (3), 334–343.

Pagliara, S., Das, R. & Palermo, M. (2008). Energy dissipation on submerged block ramps. *Journal of Irrigation and Drainage Engineering*, 134 (4), 527–532.

Pagliara, S., Palermo, M. & Carnacina, I. (2009a). Scour and hydraulic jump downstream of block ramps in expanding stilling basins. *Journal of Hydraulic Research*, 47 (4), 503–511.

Pagliara, S., Carnacina, I. & Palermo, M. (2009b). Energy dissipation in presence of block ramps with enlarged stilling basins. In: *Proceedings of the 33rd IAHR Congress, 9–14 August, Vancouver, Canada*, IAHR, pp. 3479–3486.

Pagliara, S., Palermo, M. & Lotti, I. (2009c). Sediment transport on block ramp: filling and energy recovery. *KSCE Journal of Civil Engineering*, 13 (2), 129–136.

Pagliara, S. & Palermo, M. (2010a). Influence of tailwater depth and pile position on scour downstream of block ramps. *Journal of Irrigation and Drainage Engineering*, 136 (2), 120–130.

Pagliara, S. & Palermo, M. (2010b). Energy dissipation in the stilling basin downstream of block ramps in presence of rock made sills. In: *Proceedings of the First European IAHR Congress, 4–6 May, Edinburgh, Scotland*, pp. 1–6.

Pagliara, S. & Palermo, M. (2011a). Block ramp failure mechanisms: critical discharge estimation. *Proceedings of the Institution of Civil Engineers-Water Management*, 164 (6), 303–309.

Pagliara, S. & Palermo, M. (2011b). Effect of stilling basin geometry on clear water scour morphology downstream of a block ramp. *Journal of Irrigation and Drainage Engineering*, 137 (9), 593–601.

Pagliara, S. & Palermo, M. (2011c). Discussion of the paper "Block ramp for efficient energy dissipation" by Ghare A.D., Ingle, R.N. & Porey, P.D.. *Journal of Energy Engineering*, 137 (4), 220.

Pagliara, S., Palermo, M. & Carnacina, I. (2011). Expanding pools morphology in live-bed conditions. *Acta Geophysica*, 59 (2), 296–316.

Pagliara, S. & Palermo, M. (2012). Effect of stilling basin geometry on the dissipative process in the presence of block ramps. *Journal of Irrigation and Drainage Engineering*, 138 (11), 1027–1031.

Pagliara, S., Palermo, M. & Carnacina, I. (2012). Live-bed scour downstream of block ramps for low densimetric Froude numbers. *International Journal of Sediment Research*, 27 (3), 337–350.

Pagliara, S. & Palermo, M. (2013). Rock grade control structures and stepped gabion weirs: scour analysis and flow features. *Acta Geophysica*, 61 (1), 126–150.

Palermo, M., Das, R. & Pagliara, S. (2008). Hydraulic jump classification downstream of block ramps for non-uniform channel bed material. In: S. Pagliara (ed.) *IJREWHS '08: Proceedings of the 2nd International Junior Researcher and Engineer Workshop on Hydraulic Structures, 30 July-1 August, Pisa, Italy*. Pisa, Edizioni Plus. pp. 129–134.

Peyras, L., Royet, P. & Degoutte, G. (1992). Flow and energy dissipation over stepped gabion weirs. *Journal of Hydraulic Engineering*, 118 (5), 707–717.

Platzer, G. (1983). Die Hydraulik der breiten Blochsteinrampe, rampenbeigung 1:10. Bundensanstalt, Wien, Austria.

Rajaratnam, N. & Subramanya, K. (1968). Hydraulic jumps below abrupt symmetrical expansions. *Journal of Hydraulic Division*, 94 (HY2), 481–503.

Rice, C.E., Kadavy, K.C. & Robinson, K.M. (1998). Roughness of loose rock riprap on steep slopes. *Journal of Hydraulic Engineering*, 124 (2), 179–185.

Robinson, K.M., Rice, C.E. & Kadavy, K.C. (1997). Design of rock chutes. *ASAE Paper No. 972062, St. Joseph, Michigan*.

Smith, C.D. (1989). The submerged hydraulic jump in an abrupt lateral expansion. *Journal of Hydraulic Research*, 27 (2), 257–266.

Stephenson, D. (1991). Energy dissipation down stepped spillways. *International Water Power & Dam Construction*, 43 (9), 27–30.

Whittaker, J. & Jaggi, M. (1986). *Blockschwellen*. Versuchsanstalt fur Wasserbau Hydrologie und Glaziologie, ETH, Zurich, Switzerland. Mitteilungen 91.

Chapter 3

Stepped spillways and cascades

H. Chanson[1], D.B. Bung[2] & J. Matos[3]

[1]School of Civil Engineering, The University of Queensland, Brisbane,
QLD, Australia
[2]Hydraulic Engineering Section, FH Aachen University of Applied
Sciences, Bayernallee 9, Aachen, Germany
[3]Department of Civil Engineering, Architecture and Georesources, IST,
Av. Rovisco Pais, Lisbon, Portugal

ABSTRACT

A spillway system is an aperture designed to spill safely the flood waters and dissipate the turbulent kinetic energy of the flow before it rejoins the natural river channel. The construction of
steps on the spillway may assist with the energy dissipation, thus reducing the size of the downstream stilling structure. The construction of stepped spillways is compatible with the placement methods of roller compacted concrete and gabions. The main characteristics of stepped
spillway flows are the different flow regimes depending upon the relative discharge, the high
turbulence levels and the intense flow aeration. Modern stepped spillways are characterised
by a relatively steep slope and large flow rates per unit width. The chute toe velocity may be
estimated using a graphical method, and the downstream energy dissipator must be designed
accurately with the knowledge of the air-water flow properties. As the flow patterns of stepped
spillways differ from those on smooth chutes, designers must analyse carefully stepped chute
flows and their design is far from trivial.

3.1 INTRODUCTION

Dams and weirs are man-made hydraulic structures built across a river to provide
water storage. During major rainfall events, large water inflows into the reservoir
induce a rise in the reservoir level with the risk of dam overtopping. The spillway
system is an aperture designed to spill safely the flood waters above, below or besides
the dam wall. Most small dams are equipped with an overflow structure, the spillway, which includes typically a crest, a chute and an energy dissipator at the downstream end. The energy dissipator is designed to dissipate the excess in kinetic energy
at the end of the spillway before it re-joins the natural stream. Energy dissipation
on dam spillways is achieved usually by a standard stilling basin downstream of a
steep chute in which a hydraulic jump takes place, converting the flow from supercritical to subcritical conditions, a high velocity water jet taking off from a ski jump
and impinging into a downstream plunge pool, or a plunge pool in which the chute
flow impinges and the kinetic turbulent energy is dissipated in turbulent recirculation. The construction of steps on the spillway chute may assist also with the energy
dissipation.

The stepped channel design has been used for more than 3 millenia (Knauss
1995, Chanson 2000–2001). A significant number of structures were built with a
stepped spillway systems during the 19th century through to the early 20th century

(Fig. 3.1A), when the design technique became outdated with the development of hydraulic jump stilling basin designs. Recent advances in construction materials and technology, including Roller Compacted Concrete (RCC) and polymer-coated gabion wire, led to a renewal of interest for the stepped chute design (Fig. 3.1B & 3.1C). The stepped spillway profile increases significantly the rate of energy dissipation taking place along the spillway, thus reducing the size of the downstream stilling structure. Stepped cascades are used also in water treatment plants to enhance the air-water transfer of atmospheric gases (e.g. oxygen, nitrogen) and of Volatile Organic Components (VOC).

In this chapter, the basic hydraulic characteristics of stepped spillways are reviewed. The energy dissipation and aeration characteristics are detailed, before the stepped chute design is discussed.

Figure 3.1 Stepped spillways. (A) Gold Creek dam stepped spillway (Australia), completed in 1890, on 14 April 2009 (Courtesy of Gordon Grigg) (B) Riou dam (France) on 11 February 2004 (C) Pedrógão dam (Portugal) on 4 September 2006.

3.2   HYDRAULICS OF STEPPED SPILLWAYS

3.2.1   Basic flow patterns

For a given stepped chute geometry, the flow may be a nappe flow at low flow rates, a transition flow for intermediate discharges or a skimming flow at larger flow rates. Figure 3.2 shows some photographs of nappe and skimming flows down a 26.6° stepped chute.

In the nappe flow regime, the flow progresses as a series of free-falling nappes impacting onto the downstream step (Fig. 3.2A). The energy dissipation occurs by jet breakup in air, jet impact onto the horizontal step face, and sometimes formation of a hydraulic jump on the step. Practical considerations show that the step height must be significantly larger than the critical flow depth (Fig. 3.3). For a range of intermediate flow rates, an intermediate flow pattern between nappe and skimming flow may be observed: the transition flow regime (Ohtsu and Yasuda 1997, Chanson and Toombes 2004). The transition flow is characterised by a chaotic flow motion associated with intense splashing. The transition flow pattern exhibits significant longitudinal variations of the flow properties at each step and between subsequent steps, and flow instabilities. It is recommended to avoid this flow regime for design operation at medium to large discharges. In a skimming flow regime, the water flows as a coherent stream, skimming over the pseudo-bottom formed by the step edges (Fig. 3.2B). Beneath the pseudo-bottom, cavity recirculation is maintained through the transfer of momentum from the main stream to the recirculating fluid.

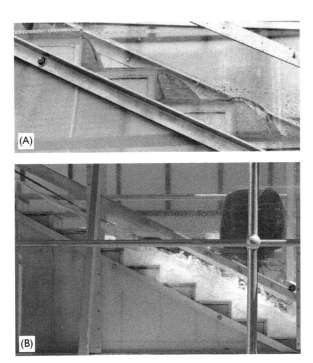

Figure 3.2   Nappe and skimming flows down a 1V:2H flat stepped chute ($h = 0.10$ m). (A) Nappe flow, $d_c/h = 0.13$. (B) Skimming flow, $d_c/h = 1.45$.

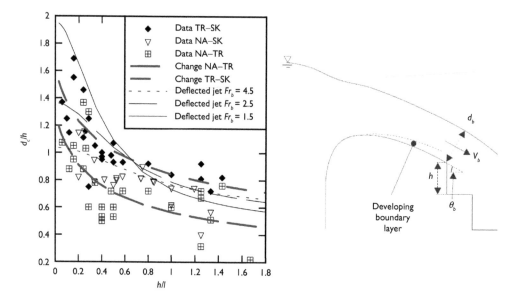

Figure 3.3 Flow conditions for the upper limit of nappe flows and lower limit of skimming flows. Comparison with older observations of transition from nappe to skimming flows. Comparison with the criterion for onset of jet deflection.

The type of flow regime is a function of the discharge and stepped configuration. The analysis of a large number of experimental observations on flat horizontal steps suggests that the upper limit of nappe flow may be approximated as

$$\left(\frac{d_c}{h}\right)_{NA-TR} = \frac{0.593}{\left(\dfrac{h}{l}+0.139\right)^{0.394}} \qquad (3.1)$$

while a lower limit of skimming flow is given by:

$$\left(\frac{d_c}{h}\right)_{TR-SK} = \frac{0.881}{\left(\dfrac{h}{l}+0.149\right)^{0.317}} \qquad (3.2)$$

where d_c is the critical flow depth, h is the vertical step height and l is the horizontal step length. Equations (3.1) and (3.2) are compared with experimental data in Figure 3.3 (thick dashed lines), together with older observations of the transitions between nappe and skimming flows. The results characterise a change in flow regime for quasi-uniform flows, and should not be applied to rapidly varied flows.

Some prototype and laboratory observations indicated the risk of jet deflection at the upstream steps if these are too high (Fig. 3.3 Right). A theoretical solution for the onset of jet deflection is (Chanson 1996):

$$\frac{d_c}{h} < \frac{Fr_b^{2/3} \times \sqrt{1 + \dfrac{1}{Fr_b^2}}}{\sqrt{1 + 2 \times Fr_b^2 \times \left(1 + \dfrac{1}{Fr_b^2}\right)^{3/2} \times \left(1 - \dfrac{\cos\theta_b}{\sqrt{1 + \dfrac{1}{Fr_b^2}}}\right)}} \tag{3.3}$$

where $Fr_b = V_b/(g \times d_b)^{1/2}$, d_b and V_b are the flow depth and mean velocity at the brink and θ_b the angle between the invert and horizontal. Equation (3.3) is shown in Figure 3.3 and may be used to prevent risks of jet deflection at the first steps, for example by installing smaller first steps.

3.2.2 Hydraulics of nappe flow

For a given stepped geometry, small discharges operate in the nappe flow regime (Fig. 3.2A). At each step brink, a free-falling nappe takes off with an air cavity and a pool of recirculating fluid underneath. Air entrainment may take place at the upper and lower nappes, with additional air being entrained by a plunging jet motion at the intersection of the lower nappe with the recirculating water. For low discharges and flat slopes, a hydraulic jump may take place downstream of the jet impact and upstream of the next step edge. For larger flow rates and steeper chutes, the flow is supercritical all along the spillway and no jump is observed.

The pressures and pressure fluctuations on the channel steps are important factors affecting the stepped spillway operation. Large hydrodynamic forces are exerted at the impact of the falling nappe, on the vertical face if the nappe is not adequately ventilated, and possibly beneath the hydraulic jump. At nappe impact, large invert pressures may be experienced next to the stagnation point, with mean stagnation pressure P_s about (Chanson 1995a):

$$\frac{P_s - P_{atm}}{\rho_w \times g \times h} = 1.25 \times \left(\frac{d_c}{h}\right)^{0.35} \tag{3.4}$$

and extreme maximum and minimum pressures of the order of magnitude:

$$(P_s)_{max} \approx P_s + 0.9 \times \rho_w \times \frac{V_i^2}{2} \tag{3.5}$$

$$(P_s)_{min} \approx P_s - 0.6 \times \rho_w \times \frac{V_i^2}{2} \tag{3.6}$$

with V_i the impact velocity of the falling nappe.

The nappe ventilation is essential for a proper spillway operation in the nappe flow regime. In absence of aeration, fluttering instabilities may develop associated with oscillations of the nappe and loud noise (Fig. 3.4 Right) (Pariset 1955). The danger to the dam structure is usually small, unless the fluttering frequency is close to the resonance frequency of the system. For a stepped spillway with flat horizontal steps, the theoretical oscillation frequencies are shown in Figure 3.4 (Left) and they are compared with experimental observations including prototype spillway data. In this section, it is assumed that the nappe cavity is adequately ventilated.

In a nappe flow, the energy dissipation occurs by jet breakup and mixing, and possibly the formation of a hydraulic jump on the horizontal step face. The total head loss along the chute ΔH equals the difference between the maximum head available H_1 and the residual head H_{res} at the downstream end of the spillway. Thus the rate of energy dissipation for an ungated stepped spillway equals (Chanson 1994a):

$$\frac{\Delta H}{H_1} = 1 - \left(\frac{0.54 \times \left(\dfrac{d_c}{h}\right)^{0.275} + \dfrac{3.43}{2} \times \left(\dfrac{d_c}{h}\right)^{-0.55}}{\dfrac{3}{2} + \dfrac{H_{dam}}{d_c}} \right) \tag{3.7}$$

where H_{dam} is the drop in elevation between the spillway crest and chute toe. Although developed for nappe flow with hydraulic jump, Equation (3.7) was tested successfully for nappe flow without hydraulic jump and stepped chute slopes up to 23° (Chanson 2001a).

A number of design criteria were proposed for stepped cascades operating with nappe flows (Binnie 1913, Stephenson 1991). These imply relatively large steps and flat slopes, a situation not often practical although suited for some river training

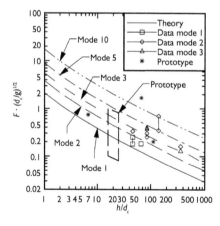

Figure 3.4 Deflected nappe oscillations. Left: comparison between theoretical calculations (Casperson 1993) and observations of nappe oscillation frequency *F*. Right: nappe oscillations at Chichester dam in April 2008: "The intake of air was quite audible, and pulsing at around five intakes per second" (Courtesy of Ken Rubeli).

and storm waterways. Practically, design engineers must size the sidewall height to prevent spray overflow. This is particularly important for embankment stepped spillways since the risk of erosion of the non-overflow section must be taken into account. A similar criterion may be used to prevent the developing spray to generate fog or ice in the surroundings in cold weather conditions.

3.2.3   Hydraulics of skimming flow

At large discharges, the water flows down a stepped spillway as a coherent stream skimming over the pseudo-bottom formed by the step edges. In the step cavities, recirculating vortices develop and the recirculation motion is maintained through the transmission of shear stress from the main flow (Fig. 3.2B). Most turbulent kinetic energy is dissipated to maintain the cavity circulation.

At the upstream end of the chute, the skimming flow free-surface is smooth and no air entrainment occurs. Once the outer edge of the developing boundary layer interacts with the free-surface, the flow is characterised by strong air entrainment. For an uncontrolled stepped spillway, the location of the inception point of free-surface aeration may be derived based upon a semi-analytical expression of the turbulent boundary layer development (Chanson 1994b,1995):

$$\frac{(L_I)_{uc}}{h \times \cos\theta} = 9.719 \times (\sin\theta)^{0.0796} \times F_*^{0.713} \tag{3.8}$$

where L_I is the streamwise distance from the crest, the subscript uc refers to an uncontrolled crest, θ is the angle between the chute slope ($\tan\theta = h/l$) and horizontal, and F_* is a Froude number defined in terms of the step roughness height:

$$F_* = \frac{q_w}{\sqrt{g \times \sin\theta \times (h \times \cos\theta)^3}} \tag{3.9}$$

A similar reasoning gives an expression of the flow depth at inception (Chanson 1994b):

$$\frac{(d_I)_{uc}}{h \times \cos\theta} = 0.4034 \times \frac{F_*^{0.592}}{(\sin\theta)^{0.04}} \tag{3.10}$$

For a gated spillway or a pressurised intake, the initial flow conditions of boundary layer development differ, and the analytical calculations imply (Chanson 2006):

$$\left(\frac{(L_I)_{pi}}{h \times \cos\theta}\right)^{1.4} = \left(\frac{(L_I)_{uc}}{h \times \cos\theta}\right)^{1.4} \times \frac{1}{1 + \dfrac{F_*^{2/3}}{\dfrac{(L_I)_{pi} \times (\sin\theta)^{1/3}}{h \times \cos\theta}} \times Fr_1^{-2/3} + \dfrac{1}{2} \times Fr_1^{4/3}} \tag{3.11}$$

$$\left(\frac{(d_I)_{pi}}{h\times\cos\theta}\right)^{1.57} = \left(\frac{(d_I)_{uc}}{h\times\cos\theta}\right)^{1.57} \times \frac{1}{1+\dfrac{\dfrac{F_*^{2/3}}{(L_I)_{pi}\times(\sin\theta)^{1/3}}\times Fr_1^{-2/3}+\dfrac{1}{2}\times Fr_1^{4/3}}{h\times\cos\theta}} \qquad (3.12)$$

where the subscript pi refers to pressurised intake conditions and Fr_1 is the intake flow Froude number. Basic considerations show that the location of the inception point of free-surface aeration is located further upstream on a controlled chute or with a pressurised intake, for an identical flow rate, slope and step height. A number of prototype and laboratory data are shown in Figure 3.5 and compared with Equations (3.8), (3.10), (3.11) and (3.12). In skimming flows on an uncontrolled crest, the prototype data followed relatively closely Equations (3.8) and (3.10).

Flow resistance in skimming flows is associated with considerable form losses and momentum transfer between the main flow and the step cavity recirculation (Rajaratnam 1990). A comprehensive re-analysis of flow resistance data in laboratory and prototype is presented in Figure 3.6, regrouping 249 data points. Figure 3.6 presents the equivalent Darcy friction factor as function of the dimensionless cavity height $h\times\cos\theta/D_H$, where D_H is the hydraulic diameter. For steep chutes ($\theta > 10°$), the data presented no obvious correlation with the relative cavity height, Reynolds and Froude numbers (Chanson et al. 2002). Overall the data compared well with a simplified analytical model of the pseudo-boundary shear stress:

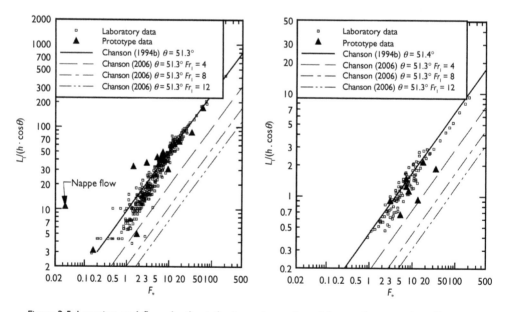

Figure 3.5 Location and flow depth at the inception point of free-surface aeration. Comparison between prototype observations (Trigomil, Dona Francisca, Pedrógão, Brushes Clough, Gold Creek, Paradise, Hinze), laboratory data, and Equations (3.8), (3.10), (3.11) and (3.12).

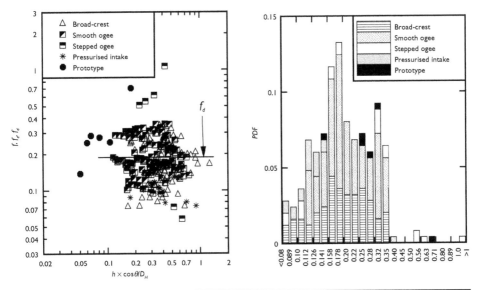

Description	Broad-crest	Smooth ogee	Ogee with steps	Pressurised intake	Prototype
Nb samples	86	121	19	17	6
Mean value	0.17	0.18	0.30	0.10	0.32
Range	0.07–0.35	0.10–0.36	0.06–1.1	0.07–0.13	0.14–0.70

Figure 3.6 Darcy-Weisbach friction factor in skimming flow above a stepped spillway as a function of the relative cavity height $h \times \cos\theta/D_H$ (Left) and its probability distribution function (Right) Comparison between laboratory ($\theta > 10°, h > 0.02$ m, $Re > 1 \times 10^5$) and prototype data.

$$f_d = \frac{2}{\sqrt{\pi}} \times \frac{1}{K} \qquad (3.13)$$

with $1/K$ being the shear layer expansion rate (Rajaratnam 1976, Schlichting 1979). Equations (3.13) thus predicts $f_d \sim 0.2$ for $K = 6$, a result comparable to experimental observations (Fig. 3.6 Left).

Altogether the skimming friction factor data appeared to be distributed around three dominant values: $f \approx 0.11, 0.17$ and 0.30 as shown in Figure 3.6 Right which presents the normalised probability distribution function of friction factor. It was suggested that flow resistance in skimming flows is not an unique function of flow rate and stepped chute geometry (Chanson 2006, Felder and Chanson 2009a). The form drag process may present several modes of excitation resulting from the vortex shedding in the shear layers downstream of each step edge. A number of laboratory data showed further that the flow properties in skimming flows oscillated between adjacent step edges (Boes 2000, Matos 2000, Chanson and Toombes 2002a, Gonzalez and Chanson 2004, Felder and Chanson 2009a). The existence of such instabilities implied that the traditional concept of 'normal flow' might not exist in skimming flows on stepped spillways.

3.3 HYDRAULIC DESIGN OF STEPPED SPILLWAYS

3.3.1 Energy dissipation

A sound estimate of the residual energy head at the stepped spillway toe is essential during the design stages. Compared to a smooth invert chute, the residual energy is drastically lower but it often requires some additional downstream energy dissipator, e.g. a stilling basin. In the skimming flow regime usually considered for stepped spillway design, a strong, energy-consuming momentum transfer between the step cavity flow and the skimming water body is observed. The residual energy head H_{res} is directly linked to the friction factor f by

$$\frac{H_{res}}{H_{max}} = \frac{\left(\dfrac{f}{8 \times \sin\theta}\right)^{1/3} + \dfrac{\alpha}{2} \times \left(\dfrac{f}{8 \times \sin\theta}\right)^{-2/3}}{\dfrac{3}{2} + \dfrac{H_{dam}}{d_c}} \qquad (3.14)$$

where $H_{max} = H_{dam} + 3/2 \times d_c$ is the maximum energy head above chute toe, θ is the chute angle and α is the kinetic energy correction factor taking into account the velocity distribution perpendicular to the pseudo-bottom (Fig. 3.8). Because of the strong turbulence, α was found to be about 1.1 (Matos 2000, Boes and Hager 2003b). The friction factor f may be taken from Figure 3.6. It must be noted that Equation (3.14) assumes quasi-uniform equilibrium flow to be achieved (Chanson 1994c). For lower dams, the residual energy must be calculated based upon the friction slope S_f (Eq. (3.16)).

Design engineers should be aware that full energy dissipation can never be achieved. A number of downstream energy dissipator designs may be considered. Commonly, standard stilling basins developed for smooth invert chutes are designed (Peterka 1958). But only few studies tested their hydraulic behavior in combination with stepped spillways (e.g. Cardoso et al. 2007, Bung et al. 2012).

3.3.2 Air entrainment

Prototype and laboratory observations highlighted the strong flow aeration on a stepped spillway. The turbulence acting next to the free-surface induces a substantial self-aeration. The flow aeration induces some flow bulking, thus requiring higher chute sidewalls, while it prevents cavitation damage. Although laboratory experiments showed the cavitation potential in skimming flows (Frizell et al. 2013), all prototype tests demonstrated an absence of cavitation damage, even for large discharges per unit width (Lin and Han 2001a), which is likely the result of air entrainment. A further application is the air-mass transfer on a stepped chute (Essery et al. 1978, Toombes and Chanson 2005).

In a skimming flow, the upstream flow is non-aerated. Downstream of the inception point of free-surface aeration, the entire water column becomes rapidly aerated as illustrated in Figure 3.7. Figure 3.7 shows some dimensionless distributions of void fraction and bubble count rate, and the data are compared with theoretical solutions of the advective diffusion equation for air bubbles and bubble count rate distributions.

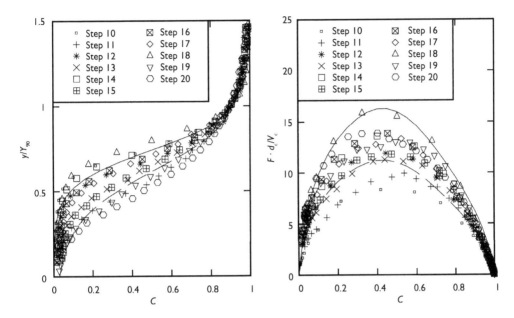

Figure 3.7 Dimensionless distributions of void fraction C (C_{mean} = 0.24 to 0.46) and bubble count rate $F \times d_c/V_c$ in skimming flows – Flow conditions: θ = 21.8°, h = 0.05 m, d_c/h = 1.75 – Comparison with an advective diffusion model for air bubbles (C_{mean} = 0.25 & 0.40, Chanson & Toombes 2002b) and bubble count rate distribution model (Toombes & Chanson 2008).

3.3.3 Design guidelines

During the last four decades, a number of stepped spillways were designed for gravity dams and embankment structures (Chanson 2001a). In the 1990s, the construction of secondary stepped spillways accounted for nearly two-thirds of dam construction in USA (Ditchey and Campbell 2000). There are several construction techniques to form a stepped slope, including gabions, reinforced earth, pre-cast concrete slabs and Roller Compacted Concrete (RCC). RCC construction and gabion placement techniques yield naturally a spillway in a simple stepped fashion. Gabion stepped chutes are usually restricted to small structures (Peyras et al. 1992), while the step face roughness and seepage flow must be accounted for (Kells 1993, Gonzalez et al. 2008, Wutrich and Chanson 2014).

The stepped spillway is typically designed to operate in a skimming flow regime. During the design process, the dam height, its downstream slope and the design discharge are known parameters. The variables include the chute width and step height, possibly the flow regime. The design engineers are often limited to select a step height within the values determined by the dam construction technique (h = 0.2 to 1.2 m with RCC). Chanson (2001b) and Gonzalez and Chanson (2007) detailed the complete design steps, and a summary follows.

At design flow, the inception point of free-surface aeration must, if possible, be located upstream of the chute downstream end to ensure that the flow is fully-developed before and minimise the residual energy at the chute toe (Fig. 3.8). The inception point characteristics may be calculated using Equations (3.8) and (3.10) for an ungated spillway and Equations (3.11) and (3.12) for a gated spillway crest.

If the spillway chute is long enough for the flow to reach uniform equilibrium, the equivalent clear-water flow depth is derived from momentum considerations:

$$d = \sqrt[3]{\frac{f \times q_w^2}{8 \times g \times \sin\theta}} \qquad (3.15)$$

where the friction factor f may be estimated from Figure 3.6 for preliminary design purposes. When normal flow conditions are not achieved before the chute toe, the flow depth may be deduced from the backwater equation:

$$\frac{\partial H}{\partial x} = -S_f = -\frac{f}{8} \times \frac{q_w^2}{g \times d^3} \qquad (3.16)$$

It must be noted that the depth-averaged air concentration C_{mean} may increase to more than 50% in the fully developed flow region. For the design of chute sidewall height, the characteristic depth $Y_{90} = d/(1 - C_{mean})$ may be more relevant and could reach twice the clear-water flow depth (Bung 2013). Some studies presented relevant design calculations to predict the longitudinal variations of d, C_{mean} and Y_{90} (Matos 2000, Chanson 2001a).

Alternatively the flow properties at the chute toe may be calculated with a simple design chart linking some well-documented theoretical considerations and experimental observations (Fig. 3.9). Developing flow and uniform equilibrium

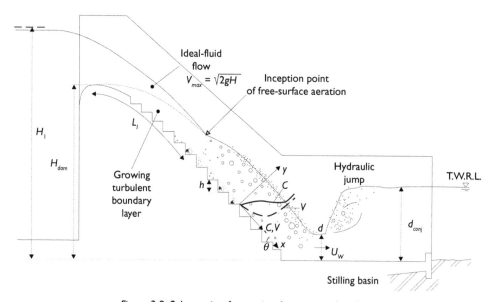

Figure 3.8 Schematic of a gravity dam stepped spillway.

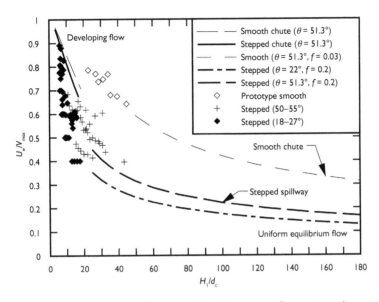

Figure 3.9 Flow velocity at the stepped spillway toe. Comparison between smooth and stepped spillway configurations.

flow calculations may be combined to provide a general trend which may be used for a preliminary design. The ideal fluid flow velocity at the downstream end of the chute is:

$$V_{max} = \sqrt{2 \times g \times (H_1 - d \times \cos\theta)} \qquad (3.17)$$

where H_1 is the upstream total head above chute toe (Fig. 3.8). The flow velocity U_w at the chute toe is smaller than the ideal fluid velocity V_{max} because of the energy losses down the stepped chute. Results are summarized in Figure 3.9 in terms of U_w/V_{max} as a function of H_1/d_c where d_c is the critical depth. Developing and uniform equilibrium flow calculations are shown for both smooth chutes and uncontrolled stepped spillways for two slopes $\theta = 22°$ (1V:2.5H) and $\theta = 51.3°$ (1V:0.8H) typical of embankment and gravity dam spillways. Prototype smooth chute data and laboratory stepped spillway data are included for comparison. Despite some scatter, Figure 3.9 provides a simple means to estimate the flow velocity U_w at the chute toe as a function of flow rate and upstream total head. Overall Figure 3.9 illustrates the slower velocity and lesser residual energy at the end of a stepped spillway compared to a smooth spillway design.

In some practical cases, for small dams and for medium dams and high unit design flow, the inception point of free-surface aeration may not take place on the chute length for the design discharge. In such cases, some simple design guidelines may be based upon analytical and semi-analytical models of the non-aerated developing flow region (Chanson 2001b, Meireles and Matos 2009).

While most modern stepped spillways consists of flat horizontal steps, recent studies suggested different step configurations that might enhance the rate of energy dissipation (André et al. 2008, Gonzalez and Chanson 2008).

Designers should be aware that the stepped spillway design is a critical process, as any failure can lead to a catastrophe. A number of key parameters should be assessed properly, including flow conveyance through the crest and chute, stepped face erosion, energy dissipation above the steps and in the downstream stilling structure, interactions between the abutments and the stepped face... In turn, some physical modelling with scaling ratios no greater than 3:1 is strongly recommended, especially if the full properties of the air-water flows are of interest, including flow bulking and air-water mass transfer. A few studies systematically investigated the aerated flow properties, at the local sub-millimetre scale, in geometrically similar models under controlled flow conditions to assess the associated-scale effects. These studies were based upon Froude and Morton similarity with undistorted models of stepped spillways (Boes and Hager 2003a, Chanson and Gonzalez 2005, Felder and Chanson 2009b). Despite the limited scope, the results demonstrated the limitations of dynamic similarity and physical modelling of highly turbulent aerated flows. They emphasized further that the selection of criteria to assess scale affects is critical and should involve a range of characteristics such as void fraction distributions, turbulence intensity distributions and distributions of bubble chords (Chanson 2009). The experimental results showed that some parameters, such as bubble sizes and turbulent scales, are likely to be affected by scale effects, even in 2:1 scale models (Felder and Chanson 2009b).

3.4 PROTOTYPE EXPERIENCE

3.4.1 Prototype tests

A number of tests on prototype stepped spillway were performed in China, Russia, Germany, UK and Africa. Some extensive tests were performed on the Dachaoshan RRC dam spillway in 2002 with discharge per unit width up to 72 m²/s (Lin and Han 2001a,b). Detailed inspections after each series of test indicated no damage nor any sign of cavitation pitting (K. Lin 2002,2012 Pers. Comm.). A series of tests conducted at the Dneiper hydroplant were performed with discharges per unit width up to 59.4 m²/s, velocities up to 23 m/s, a step height of 0.405 m and water depths between 0.5 to 3 m (Grinchuk et al. 1977). The stability of the wedge-shaped blocks and the flow resistance was successfully tested and some flow resistance data are reported in Figure 3.6. No sign of damage and cavitation pitting was reported. Some field testing was conducted successfully on the Brushes Clough dam spillway up to 1 m²/s, although for a short duration (Baker 2000). The data highlighted the substantial flow aeration. Some inspection of the M'Bali RCC dam stepped spillway after a wet season operation showed no sign of damage and pitting (Bindo et al. 1993).

During an uncontrolled spillway release on 2 November 1998, some pseudo-periodic, self-sustained unstabilities developed along the Sorpe dam pooled stepped spillway, the surging waters overtopping the chute sidewalls and causing a hazard to nearby tourists (Chanson 2001a, Thorwarth 2008). Some full-scale tests were

Figure 3.10 Prototype stepped spillway operations. (A) Hinze dam spillway ($\theta = 51.3°$, $h = 1.2$ m, $Q = 200$ m³/s, $d_c/h = 2.5$ on 29 Jan. 2013). (B1) Paradise dam spillway ($\theta = 57.4°$, $h = 0.62$ m, $Q = 2,320$ m³/s, $d_c/h = 2.85$ on 5 Mar. 2013). (B2) Paradise dam spillway ($\theta = 57.4°$, $h = 0.62$ m, $Q = 6,000$ m³/s, $d_c/h = 5.4$ on 30 Dec. 2010). (C) Pedrógão dam ($\theta = 53.1°$, $h = 0.6$ m, on 10 Jan. 2010) (Courtesy of EDIA). (D) Dona Francisca dam ($\theta = 53.1°$, $h = 0.6$ m, in July 2011) (Courtesy of Marcelo Marques).

undertaken in 2002–2003 to investigate the jump wave occurrence and properties (P. Kamrath 2003 Pers. Comm.).

3.4.2 Discussion: Operation, experience, accidents

A number of prototype overflow events were documented worldwide (Chanson 1995a,2001a). In China, the Shuidong Hydropower Station experienced a peak flow of 90 m²/s (He and Zeng 1995). The RCC dam spillway was protected by conventional concrete steps ($h = 0.9$ m) with a 8-cm chamfer. A movie documentary showed no abnormal operation and inspections after the flood event indicated no sign of damage. In Queensland (Australia), the operational record of several overflow stepped weirs demonstrated the soundness of the timber crib piled weir design. One structure, Cunningham weir, was overtopped for more than 2 months in 1956, with a maximum discharge per unit width in excess of 60 m²/s. Only minor damage was experienced and the weir is still operational today. Between 2010 and 2013, several stepped spillways operated during a succession of major flood events. Figure 3.10 illustrates the operation of some structures. No damage to the stepped chute itself was reported despite some exceptional flood events lasting for weeks, although some scour of the Paradise dam stilling basin was documented. In two cases, the stepped spillway operated when the dam was overtopped during construction: Trigomil dam (Mexico, 1992), Cotter dam (Australia, 2012). In each case the damage was small, including for the unprotected Roller Compacted Concrete (RCC) steps. These examples are only a few and there are ample documentations on stepped spillway operations, including historical records, in Europe, Africa, America, Asia and Australia.

In hydraulic engineering, there is no better proof of design soundness than a successful operational record. This is particularly true with stepped spillway structures, in use for more than 3,500 years (Chanson 1995b,2000–2001). A number of field testings of stepped spillway structures showed a sound operation of the prototype spillways with discharges per unit width up to 72 m²/s. These investigations are complemented by a large number of prototype experiences with stepped spillway operation during major to exceptional floods. For example, Figure 3.1A shows a 1890 staircase spillway with 1.4 m high steps which has successfully operated for over 120 years, including during large flood events. Relevant reviews of stepped spillway operation include Chanson (1995a,2001a). All the observations indicated an absence of cavitation pitting and damage to the steps. The long-lasting successful operation, for more than 3,000 years, highlights the design soundness of stepped spillways, while emphasising the importance of expert hydraulic engineering during the design stages.

3.5 CONCLUSION

The Greek hydraulic engineers were probably the first to design overflow dams with stepped spillways, more than 3,500 years ago. The overflow stepped spillways are selected to contribute to the stability of the dam, for their simplicity of shape and to reduce flow velocities. The steps increase significantly the rate of energy dissipation taking place on the steep chute and reduce the size of the required downstream energy dissipation system and the risks of scouring. The construction of stepped spillway is compatible with the slipforming and placement methods of RCC and with the construction techniques of gabion weirs. The main characteristics of stepped spillway flows are the different flow regimes (nappe, transition, skimming) depending upon the relative discharge, the high turbulence levels and the intense flow aeration.

Modern stepped spillways are characterised by a relatively steep slope and large flow rates per unit width for which the flow spills as a skimming flow. The chute toe velocity may be estimated using a graphical method (Fig. 3.9), and the downstream energy dissipator may be designed more accurately with the knowledge of the air-water flow properties. As the flow patterns of stepped spillways differ from those on smooth chutes, designers must analyse carefully stepped chute flows. The design is far from trivial. Current expertise is focused on the hydraulics of skimming flow on prismatic rectangular channels with flat horizontal steps. Little information is available for other geometries.

ACKNOWLEDGEMENTS

The authors thank their students, former students and co-workers. They also thank all the people who provided them with relevant informations. The first author acknowledges the financial support of the Australian Research Council (Grants DP0878922 & DP120100481).

REFERENCES

André, S., Bollaert, J.L., and Schleiss, A. (2008). Ecoulements Aérés sur Evacuateurs en Marches d'Escalier Equipées de Macro-Rugosités – Partie 1: Caractéristiques Hydrauliques. *La Houille Blanche*, (1), 91–100 (in French).

Baker, R. (2000). Field Testing of Brushes Clough Stepped Block Spillway. In Minor, H.E. & Hager, W.H. (eds.) *Proceedings International Workshop on Hydraulics of Stepped Spillways*, Zürich, Switzerland, Balkema Publisher, 13–20.

Bindo, M., Gautier, J., and Lacroix, F. (1993). The Stepped Spillway of M'Bali Dam. *International Water Power and Dam Construction*, 45 (1), 35–36.

Binnie, A.R. (1913). *Rainfall Reservoirs and Water Supply*. Constable & Co, London, UK, 157 pages.

Boes, R.M. (2000). Zweiphasenströmung und Energieumsetzung an Grosskaskaden. (Two-Phase Flow and Energy Dissipation on Cascades.) *Ph.D. thesis*, VAW-ETH, Zürich, Switzerland (in German).

Boes, R.M., and Hager, W.H. (2003a). Two-phase flow characteristics of stepped spillways. *Journal of Hydraulic Engineering*, ASCE, 129 (9), 661–670.

Boes, R.M., and Hager, W.H. (2003b). Hydraulic design of stepped spillways. *Journal of Hydraulic Engineering*, ASCE, 129 (9), 671–679.

Bung, D.B. (2013). Non-intrusive detection of air-water surface roughness in self-aerated chute flows, *Journal of Hydraulic Research*, 51 (3), 322–329

Bung, D.B., Sun, Q., Meireles, I., Matos, J., and Viseu, T. (2012). USBR type III stilling basin performance for steep stepped spillways, *Proceedings of the 4th IAHR International Symposium on Hydraulic Structures*, APRH – Associação Portuguesa dos Recursos Hídricos, J. Matos, S. Pagliara & I. Meireles Eds., 9–11 February 2012, Porto, Portugal, Paper 4, 8 pages (CD-ROM).

Cardoso, G., Meireles, I., and Matos, J. (2007). Pressure head along baffle stilling basins downstream of steeply sloping stepped chutes. *Proceedings of 32nd IAHR Biennial Congress*, Venice, Italy, G. Di Silvio and S. Lanzoni Editors, 10 pages (CD-ROM).

Casperson, L.W. (1993). Fluttering Fountains. *Journal of Sound and Vibrations*, 162 (2), 251–262.

Chanson, H. (1994a). Hydraulics of Nappe Flow Regime above Stepped Chutes and Spillways. *Australian Civil Engineering Transactions*, I.E.Aust., CE36 (1), 69–76.

Chanson, H. (1994b). Hydraulics of Skimming Flows over Stepped Channels and Spillways. *Journal of Hydraulic Research*, IAHR, 32 (3), 445–460.

Chanson, H. (1994c). Comparison of Energy Dissipation between Nappe and Skimming Flow Regimes on Stepped Chutes. *Journal of Hydraulic Research*, IAHR, 32 (2), 213–218. Errata: 33 (1), 113.

Chanson, H. (1995a). *Hydraulic Design of Stepped Cascades, Channels, Weirs and Spillways*. Pergamon, Oxford, UK.

Chanson, H. (1995b). History of Stepped Channels and Spillways: a Rediscovery of the 'Wheel'. *Canadian Journal of Civil Engineering*, 22 (2), 247–259.

Chanson, H. (1996). Prediction of the Transition Nappe/Skimming Flow on a Stepped Channel. *Journal of Hydraulic Research*, IAHR, 34 (3), 421–429.

Chanson, H. (2000–2001). Historical Development of Stepped Cascades for the Dissipation of Hydraulic Energy. *Transactions of the Newcomen Society*, 71 (2), 295–318.

Chanson, H. (2001a). *The Hydraulics of Stepped Chutes and Spillways*. Balkema, Lisse, The Netherlands.

Chanson, H. (2001b). Hydraulic Design of Stepped Spillways and Downstream Energy Dissipators. *Dam Engineering*, 11 (4), 205–242.

Chanson, H. (2006). Hydraulics of Skimming Flows on Stepped Chutes: the Effects of Inflow Conditions? *Journal of Hydraulic Research*, IAHR, 44 (1), 51–60.

Chanson, H. (2009). Turbulent Air-water Flows in Hydraulic Structures: Dynamic Similarity and Scale Effects. *Environmental Fluid Mechanics*, 9 (2), 125–142 (DOI: 10.1007/s10652-008-9078-3).

Chanson, H., and Gonzalez, C.A. (2005). Physical modelling and scale effects of air-water flows on stepped spillways. *Journal of Zhejiang University SCIENCE*, 6A (3), 243–250.

Chanson, H., and Toombes, L. (2002a). Experimental Study of Gas-Liquid Interfacial Properties in a Stepped Cascade Flow. *Environmental Fluid Mechanics*, 2 (3), (DOI: 10.1023/A:1019884101405).

Chanson, H., and Toombes, L. (2002b). Air-Water Flows down Stepped Chutes: Turbulence and Flow Structure Observations. *International Journal of Multiphase Flow*, 27 (11), 1737–1761.

Chanson, H., and Toombes, L. (2004). Hydraulics of Stepped Chutes: the Transition Flow. *Journal of Hydraulic Research*, IAHR, 42 (1), 43–54.

Chanson, H., Yasuda, Y., and Ohtsu, I. (2002). Flow Resistance in Skimming Flows and its Modelling. *Canadian Journal of Civil Engineering*, 29 (6), 809–819.

Ditchey, E.J., and Campbell, D.B. (2000). Roller Compacted Concrete and Stepped Spillways. In Minor, H.E. & Hager, W.H. (eds.) *Proceedings International Workshop on Hydraulics of Stepped Spillways*, Zürich, Switzerland, Balkema Publisher, 171–178.

Essery, I.T.S., Tebbutt, T.H.Y., and Rasaratnam, S.K. (1978). *Design of spillways for reaeration of polluted waters*. CIRIA Report No. 72, January, London.

Felder, S., and Chanson, H. (2009a). Energy Dissipation, Flow Resistance and Gas-Liquid Interfacial Area in Skimming Flows on Moderate-Slope Stepped Spillways. *Environmental Fluid Mechanics*, 9 (4), 427–441 (DOI: 10.1007/s10652-009-9130-y).

Felder, S., and Chanson, H. (2009b). Turbulence, Dynamic Similarity and Scale Effects in High-Velocity Free-Surface Flows above a Stepped Chute. *Experiments in Fluids*, 47 (10), 1–18 (DOI: 10.1007/s00348-009-0628-3).

Frizell, K.W., Renna, F.M., and Matos, J. (2013). Cavitation Potential of Flow on Stepped Spillways. *Journal of Hydraulic Engineering*, ASCE, 139 (6), 630–636 (DOI: 10.1061/(ASCE)HY.1943-7900.0000715).

Gonzalez, C.A., and Chanson, H. (2004). Interactions between Cavity Flow and Main Stream Skimming Flows: an Experimental Study. *Canadian Journal of Civil Engineering*, 31 (1), 33–44.

Gonzalez, C.A., and Chanson, H. (2007). Hydraulic Design of Stepped Spillways and Downstream Energy Dissipators for Embankment Dams. *Dam Engineering*, 17 (4), 223–244.

Gonzalez, C.A., and Chanson, H. (2008). Turbulence and Cavity Recirculation in Air-Water Skimming Flows on a Stepped Spillway. *Journal of Hydraulic Research*, IAHR, 46 (1), 65–72.

Gonzalez, C.A., Takahashi, M., and Chanson, H. (2008). An Experimental Study of Effects of Step Roughness in Skimming Flows on Stepped Chutes. *Journal of Hydraulic Research*, IAHR, 46 (Extra Issue 1), 24–35.

Grinchuk, A.S., Pravdivets, Y.P., and Shekhtman, N.V. (1977). Test of Earth Slope Revetments Permitting Flow of Water at Large Specific Discharges. *Gidrotekhnicheskoe Stroitel'stvo*, (4), 22–26 (in Russian).

He, G., and Zeng, X. (1995). The Integral RCC Dam Design Characteristics and Optimization Design of its Energy Dissipator in Shuidong Hydropower Station. *Proceedings International Symposium on RCC Dams*, Santander, Spain, IECA-CNEGP, Vol. 1, 405–412.

Kells, J.A. (1993). Spatially Varied Flow over Rockfill Embankments. *Canadian Journal of Civil Engineering*, 20, 820–827.

Knauss, J. (1995). ΤΗΣ ΓΡΙΑΣ ΤΟ ΠΗΔΗΜΑ, der Altweibersprung. Die rätselhafte alte Talsperre in der Glosses-Schlucht bei Alyzeia in Akarnanien. *Archäologischer Anzeiger*, 5, 138–162 (in German).

Lin, K., and Han, L (2001a). Stepped Spillway for Dachaoshan RCC Dam. In Burgi, P.H. & Gao, J. (eds.) *Proceedings 29th IAHR Biennial Congress Special Seminar*, Beijing, China, SS2 Key Hydraulics Issues of Huge Water Projects, 88–93.

Lin, K., and Han, L. (2001b). Stepped Spillway for Dachaoshan RCC Dam. *Shuili Xuebao*, Beijing, China, Special Issue IAHR Congress (9), 84–87 (in Chinese).

Matos, J. (2000). Hydraulic Design of Stepped Spillways over RCC Dams. In Minor, H.E. & Hager, W.H. (eds.) *Proceedings International Workshop on Hydraulics of Stepped Spillways*, Zürich, Switzerland, Balkema Publisher, 187–194.

Meireles, I., and Matos, J. (2009). Skimming flow in the nonaerated region of stepped spillways over embankment dams. Journal of Hydraulic Engineering, ASCE, 135 (8), 685–689.

Ohtsu, I., and Yasuda, Y. (1997). Characteristics of Flow Conditions on Stepped Channels. *Proceedings of 27th IAHR Biennial Congress*, San Francisco, USA, Theme D, 583–588.

Pariset, E. (1955). Etude sur la Vibration des Lames Déversantes. (Study of the Vibration of Free-Falling Nappes) *Proceedings 6th IAHR Biennial Congress*, The Hague, The Netherlands, Vol. 3, paper C21, 1–15.

Peterka, A.J. (1958). Hydraulic Design of Stilling Basins and Energy Dissipators, *Bureau of Reclamation*, U.S. Department of the Interior, Denver.

Peyras, L., Royet, P., and Degoutte, G. (1992). Flow and Energy Dissipation over Stepped Gabion Weirs. Journal of Hydraulic Engineering, ASCE, 118 (5), 707–717.

Rajaratnam, N. (1990). Skimming Flow in Stepped Spillways. *Journal of Hydraulic Engineering*, ASCE, 116 (4), 587–591.

Schlichting, H. (1979). *Boundary Layer Theory*. McGraw-Hill, New York, USA, 7th edition.

Thorwarth, J. (2008). Hydraulisches Verhalten von Treppengerinnen mit eingetieften Stufen – Selbstinduzierte Abflussinstationaritäten und Energiedissipation. (Hydraulics of Pooled Stepped Spillways – Self-induced Unsteady Flow and Energy Dissipation.) *Ph.D. thesis*, University of Aachen, Germany. (in German).

Toombes, L., and Chanson, H. (2005). Air-Water Mass Transfer on a Stepped Waterway. *Journal of Environmental Engineering*, ASCE, 131 (10), 1377–1386.

Toombes, L., and Chanson, H. (2008). Interfacial Aeration and Bubble Count Rate Distributions in a Supercritical Flow Past a Backward-Facing Step. *International Journal of Multiphase Flow*, 34 (5), 427–436 (doi.org/10.1016/j.ijmultiphaseflow.2008.01.005).

Wutrich, D., and Chanson, H. (2014). Hydraulics, air entrainment and energy dissipation on gabion stepped weir. *Journal of Hydraulic Engineering*, ASCE, 140 (9), 04014046, 10 pages (DOI: 10.1061/(ASCE)HY.1943-7900.0000919).

Chapter 4

Hydraulic jumps and stilling basins

H. Chanson[1] & R. Carvalho[2]

[1]School of Civil Engineering, The University of Queensland, Brisbane, QLD, Australia
[2]Department of Civil Engineering, University of Coimbra, Coimbra, Portugal

ABSTRACT

A hydraulic jump is a rapid transition from a high-velocity open channel flow to a slower fluvial motion. It is commonly experienced in streams and rivers, in industrial channels, during manufacturing processes and downstream of dam spillways. The air is entrapped at the jump toe that is a discontinuity between the impinging flow and the roller. The impingement perimeter is a source of vorticity. The air-water shear layer is characterised by a transfer of momentum from the high-velocity jet flow to the recirculation region above, as well as by an advective diffusion of entrained air bubbles. Herein the theoretical and experimental modelling of hydraulic jumps is presented with a focus on the two-phase flow properties. Later the hydraulic design of hydraulic jump stilling basin is developed and prototype experiences are discussed.

4.1 INTRODUCTION

In an open channel, a hydraulic jump is the sudden and rapid transition from a supercritical to subcritical flow. The transition is an extremely turbulent flow associated with the development of large-scale turbulence, surface waves and spray, energy dissipation and air entrainment, and it is characterised by strong dissipative processes (Fig. 4.1). Figure 4.1A illustrates a hydraulic jump in a culvert inlet during a flash flood. Figure 4.1B shows a hydraulic jump downstream of a dam spillway during a major flood. The rate of energy dissipation in the stilling basin was close to 2.4 GW. Figure 4.1C presents a smaller hydraulic jump during some inland flooding in South-East Queensland (Australia). In each case, the flow is highly turbulent and the white colour of the waters highlights the strong air bubble entrainment in the natural flow.

The turbulent flow in a hydraulic jump is extremely complicated, and it remains a challenge to scientists and researchers (Rajaratnam 1967, Chanson 2009a). Some basic features of turbulent jumps include the turbulent flow motion with the development of large-scale vortices, the air bubble entrapment at the jump toe and the intense interactions between entrained air and coherent turbulent structures in the hydraulic jump roller, for example seen in Figure 4.1. To date turbulence measurements in hydraulic jumps are limited, but for some pioneering studies (Rouse et al. 1959, Resch and Leutheusser 1972).

This section reviews the progress and development in the understanding of turbulence and air-water flow properties of hydraulic jumps, before reviewing the hydraulic design of hydraulic jump stilling basins. The focus is on the turbulent hydraulic jump

(A, Left) Hydraulic jump in a culvert inlet along Norman Creek during a flash flood in Brisbane (QLD, Australia) on 20 May 2009 – Flow from left to right
(B, Right) Hydraulic jump energy dissipator downstream of the Paradise dam stepped spillway (Australia) on 30 December 2010 on the Burnett River (QLD, Australia) – Looking upstream at the jump roller, $Q \approx 6{,}000$ m^3/s

(C) Hydraulic jump during Black Snake Creek flood, Marburg (QLD, Australia) on 11 January 2011 – Flow from foreground left to background left, looking downstream

Figure 4.1 Hydraulic jumps in civil engineering applications.

with a marked roller operating with high-Reynolds numbers. These flow conditions are typical of civil and environmental engineering applications including energy dissipators at spillway toe (Fig. 4.1). These hydraulic jumps are characterised by some complicated turbulent air-water flow features.

4.2 BASIC THEORY

4.2.1 Momentum integral treatment

A hydraulic jump is characterised by a sudden rise in free-surface elevation (Figs. 4.1 and 4.2) and a discontinuity of the pressure and velocity fields. In an integral form, the continuity and momentum principles give a system of equations linking the flow properties upstream and downstream of the jump (Lighthill 1978, Chanson 2012):

$$Q = V_1 \times A_1 = V_2 \times A_2 \tag{4.1}$$

$$\rho \times Q \times (\beta_2 \times V_2 - \beta_1 \times V_1) = \iint_{A_1} P \times dA - \iint_{A_2} P \times dA - F_{fric} + W \times \sin\theta \tag{4.2}$$

where Q is the water discharge, d and V are the flow depth and velocity respectively, ρ is the water density, g is the gravity acceleration, A is the flow cross-sectional area, β is a momentum correction coefficient, P is the pressure, the subscripts 1 and 2 refer to the upstream and downstream flow conditions respectively (Fig. 4.2), F_{fric} is the flow resistance force, W is the weight force and θ is the angle between the bed slope and horizontal. Equations (4.1) and (4.2) are valid for a stationary hydraulic jump. Neglecting flow resistance ($F_{fric} = 0$), assuming $\beta_1 = \beta_2 = 1$ and for a horizontal bed ($\theta = 0$), Equation (4.2) yields:

$$\rho \times Q \times (V_2 - V_1) = \iint_{A_1} P \times dA - \iint_{A2} P \times dA \tag{4.3}$$

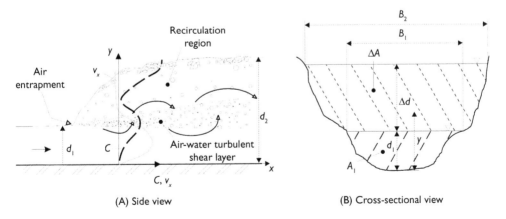

(A) Side view   (B) Cross-sectional view

Figure 4.2 Definition sketch of a hydraulic jump with a marked roller.

The difference in pressure forces may be expressed analytically assuming hydrostatic pressures upstream and downstream of the jump. The resultant consists of a pressure increase $\rho \times g \times (d_2 - d_1)$ applied to the area A_1 and the pressure force acting on the area $(A_2 - A_1)$:

$$\int_{A_1}^{A_2} \int \rho \times g \times (d_2 - y) \times dA = \frac{1}{2} \times \rho \times g \times (d_2 - d_1)^2 \times B' \tag{4.4}$$

where y is the distance normal to the bed, d is the flow depth, and B' is a characteristic free-surface width ($B_1 < B' < B_2$) with B_1 and B_2 the upstream and downstream free-surface widths (Fig. 4.2). Note that another free-surface width B may be defined as:

$$\int_{A_1}^{A_2} \int dA = A_2 - A_1 = (d_2 - d_1) \times B \tag{4.5}$$

The combination of continuity and momentum principles gives a series of expressions:

$$V_1^2 = \frac{1}{2} \times \frac{g \times A_2}{A_1 \times B} \times \left(\left(2 - \frac{B'}{B} \right) \times A_1 + \frac{B'}{B} \times A_2 \right) \tag{4.6}$$

$$(V_1 - V_2)^2 = \frac{1}{2} \times \frac{g \times (A_2 - A_1)^2}{B \times A_1 \times A_2} \times \left(\left(2 - \frac{B'}{B} \right) \times A_1 + \frac{B'}{B} \times A_2 \right) \tag{4.7}$$

In dimensionless terms, it yields:

$$\frac{V_1^2}{g \times \dfrac{A_1}{B_1}} = \frac{1}{2} \times \frac{A_2}{A_1} \times \frac{B_1}{B} \times \left(\left(2 - \frac{B'}{B} \right) + \frac{B'}{B} \times \frac{A_2}{A_1} \right) \tag{4.8}$$

Equation (4.8) provides an analytical solution of the square of Froude number as a function of the ratios A_2/A_1, B'/B and B_1/B. Note the definition of the Froude number $Fr_1 = V_1/\sqrt{g \times A_1/B_1}$ which was developed based upon momentum considerations and is identical to an expression derived from energy considerations (Henderson 1966). Equation (4.8) may be expressed as the ratio of conjugate depths A_2/A_1 being a function of the upstream Froude number:

$$\frac{A_2}{A_1} = \frac{1}{2} \times \frac{\sqrt{\left(2 - \dfrac{B'}{B} \right)^2 + \dfrac{8 \times B'}{B_1} \times Fr_1^2} - \left(2 - \dfrac{B'}{B} \right)}{\dfrac{B'}{B}} \tag{4.9}$$

Equation (4.9) is valid for a hydraulic jump in an irregular channel and it does account for the effects of the channel cross-sectional shape. When the approximation $B = B' = B_1$ holds (e.g. U-shape channels), Equation (4.9) becomes:

$$\frac{A_2}{A_1} = \frac{1}{2}\left(\sqrt{1 + 8 \times Fr_1^2} - 1\right) \tag{4.10}$$

For a rectangular channel, Equation (4.10) becomes the Bélanger equation (Bélanger 1841, Chanson 2009b):

$$\frac{d_2}{d_1} = \frac{1}{2}\left(\sqrt{1 + 8 \times Fr_1^2} - 1\right) \tag{4.11}$$

and Fr_1 becomes: $Fr_1 = V_1/\sqrt{g \times d_1}$.

In presence of flow resistance, the equations of conservation of momentum may be derived for a flat horizontal channel and the combination yields:

$$V_1^2 = \frac{1}{2} \times \frac{g \times A_2}{A_1 \times B} \times \left(\left(2 - \frac{B'}{B}\right) \times A_1 + \frac{B'}{B} \times A_2\right) + \frac{A_2}{A_2 - A_1} \times \frac{F_{fric}}{\rho \times A_1} \tag{4.12}$$

$$(V_1 - V_2)^2 = \frac{1}{2} \times \frac{g \times (A_2 - A_1)^2}{B \times A_1 \times A_2} \times \left(\left(2 - \frac{B'}{B}\right) \times A_1 + \frac{B'}{B} \times A_2\right) + \frac{A_2}{A_2 - A_1} \times \frac{F_{fric}}{\rho \times g \times \dfrac{A_1^2}{B}} \tag{4.13}$$

After re-arranging a dimensionless expression is:

$$Fr_1^2 = \frac{1}{2} \times \frac{A_2}{A_1} \times \frac{B_1}{B} \times \left(\left(2 - \frac{B'}{B}\right) + \frac{B'}{B} \times \frac{A_2}{A_1}\right) + \frac{A_2}{A_2 - A_1} \times \frac{F_{fric}}{\rho \times g \times \dfrac{A_1^2}{B}} \tag{4.14}$$

Equation (4.14) provides a solution in terms of the upstream Froude number as a function of the ratio of conjugate cross-sectional areas, the flow resistance force and the irregular cross-sectional shape properties. The effects of bed friction on hydraulic jump properties were tested on irregular channels (Chanson 2012). Figure 4.3 presents the upstream Froude number as a function of ratio of the conjugate cross-section areas A_2/A_1 for two applications. For a fixed upstream Froude number, the results imply a smaller ratio of conjugate depths d_2/d_1 with increasing flow resistance (Fig. 4.3). The finding is consistent with physical data in laboratory flumes (Leutheusser and Schiller 1975, Pagliara et al. 2008). The effects of flow resistance decrease with increasing upstream Froude number, becoming negligible for Froude numbers greater than 2 to 3 depending upon the channel shape (Fig. 4.3).

Considering a hydraulic jump down a smooth sloping rectangular prismatic channel, the combination of the continuity and momentum principles gives a physically

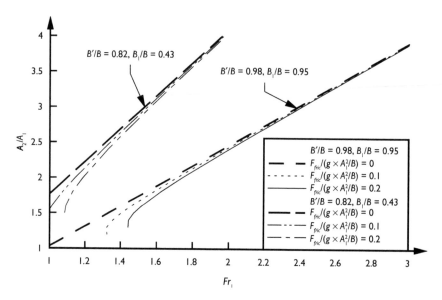

Figure 4.3 Effects of the flow resistance on the solution of the momentum equation (Eq. (4.14)) – Application to two test cases: $B'/B = 0.98$ and $B_1/B = 0.95$ (lower lines), and $B'/B = 0.82$ and $B_1/B = 0.43$ (upper lines).

meaningful solution assuming $\cos\theta \approx 1$, where θ is the angle between the bed slope and horizontal, positive for a channel sloping downwards in the downstream direction. Namely the ratio of conjugate depths equals:

$$\frac{d_2}{d_1} = \frac{1}{2} \times \left(\sqrt{(1-\varepsilon)^2 + 8 \times \frac{Fr_1^2}{1-\varepsilon}} - (1-\varepsilon) \right) \tag{4.15}$$

where ε is a dimensionless coefficient defined as:

$$\varepsilon = \frac{Vol \times \sin\theta}{B_1 \times d_1^2 \times (Fr_1^2 - 1)} \tag{4.16}$$

with Vol the volume of the control volume encompassing the bore front, such as $W = \rho \times g \times Vol$. Equation (4.16) implies that the ratio of conjugate depths increases with increasing bed slope for a given Froude number.

4.2.2 Turbulent shear flow

A key feature of a hydraulic jump with marked roller is the developing shear layer and the recirculation region above (Fig. 4.2A). The turbulent shear flow is somehow analogous to a wall jet. Assuming a two-dimensional incompressible steady flow in which the longitudinal velocity component is much greater than the vertical velocity component, the differential form of the equations of continuity and momentum becomes:

$$\frac{\partial v_x}{\partial x} + \frac{\partial v_y}{\partial y} = 0 \tag{4.17}$$

$$v_x \times \frac{\partial v_x}{\partial x} + v_y \times \frac{\partial v_x}{\partial y} = -\frac{1}{\rho} \times \frac{\partial P}{\partial x} + \frac{1}{\rho} \times \frac{\partial \tau_t}{\partial y} \tag{4.18}$$

where x and y are respectively the longitudinal and vertical directions, the subscripts x and y refers to the components in these directions, v is the local time-averaged velocity component and τ_t is the turbulent stress. Equation (4.16) is developed along the longitudinal flow direction assuming a horizontal channel and the turbulent stress is assumed much greater than the laminar stress (Rajaratnam 1976, George et al. 2000). For an ideal wall jet (Fig. 4.4, Left), the boundary conditions are:

$$v_x(y = 0) = v_x(y = +\infty) = 0 \tag{4.19a}$$

$$\int_{y=0}^{+\infty} \rho \times v_x^2 \times dy = constant \tag{4.19b}$$

where the second boundary condition states the conservation of the longitudinal momentum flux (per unit width) in absence of longitudinal pressure gradient.

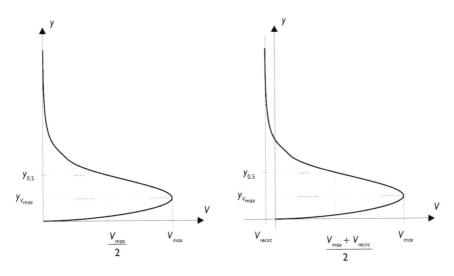

Figure 4.4 Velocity profiles in an ideal wall jet (Left) and in a hydraulic jump (Right).

Dimensional considerations and physical data implied (Wygnanski et al. 1992, Eriksson et al. 1998, George et al. 2000):

$$\frac{v_x}{V_{max}} = \left(\frac{y}{y_{V_{max}}}\right)^{1/N} \qquad 0 < y < y_{V_{max}} \qquad (4.20a)$$

$$\frac{v_x}{V_{max}} = \exp\left(-0.88 \times \left(\frac{y - y_{V_{max}}}{y_{0.5}}\right)^2\right) \qquad y_{V_{max}} < y \qquad (4.20b)$$

where V_{max} is the maximum wall jet velocity at a distance x from the nozzle, $y_{V_{max}}$ is the elevation where $v_x = V_{max}$ and $y_{0.5}$ is the elevation where $v_x = V_{max}/2$ (Fig. 4.4, Left). The longitudinal variations in maximum velocity V_{max} and characteristic height $y_{0.5}$ follow:

$$\frac{V_{max}}{V_1} \propto \left(\frac{y_{0.5}}{d_1}\right)^{1/n} \qquad (4.21a)$$

$$\frac{y_{0.5}}{d_1} \propto \frac{x - x_o}{d_1} \qquad (4.21b)$$

in which d_1 is the wall jet nozzle height, V_1 is the nozzle velocity, and x_o is the virtual origin of the boundary shear flow growth typically located upstream of the nozzle (Schwarz and Cosart 1964, Wygnanski et al. 1992). Experimental observations in smooth channels (Eriksson et al. 1998, George et al. 2000) implied $n \approx -2$ and

$$\frac{y_{0.5}}{d_1} = 0.078 \times \frac{x + 4.23}{d_1} \qquad (4.21c)$$

It is noteworthy that the above results (Eq. (4.18) & (4.19)) are general and observed for both smooth and rough surfaces (Tachie et al. 2004, Rostamy et al. 2009).

In a hydraulic jump, the basic differences with the ideal wall jet are the effect of longitudinal pressure gradient $\partial P/\partial x < 0$ and the equation of conservation of mass. Another difference is the natural forcing which is likely to enhance the two-dimensionality and periodicity of coherent structures as shown by Katz et al. (1992). For a rectangular channel, the boundary conditions are:

$$\int_{y=0}^{d} v_x \times dy = q \qquad (4.22a)$$

$$\int_{y=0}^{d} \rho \times v_x \times \frac{\partial v_x}{\partial x} \times dy = -\frac{\partial}{\partial x}\left(\int_{y=0}^{d} P \times dy\right) \qquad (4.22b)$$

where q is the water discharge per unit width and d is the local flow depth. Equation (4.20a) implies the existence of a recirculation region in the upper flow region as sketched in Figures 4.2 (Left) and 4.4 (Right). Assuming a hydrostatic pressure distribution, Equation (4.20b) becomes:

$$\int_{y=0}^{d} v_x \times \frac{\partial v_x}{\partial x} \times dx = -g \times d \times \frac{\partial d}{\partial x} \qquad (4.22c)$$

For hydraulic jumps in horizontal rectangular channels, physical data yielded (Chanson and Brattberg 2000, Chanson 2010, Chachereau and Chanson 2011b):

$$\frac{v_x}{V_{max}} = \left(\frac{y}{y_{V_{max}}}\right)^{1/N} \qquad\qquad 0 < y < y_{V_{max}} \qquad (4.23a)$$

$$\frac{v_x - V_{recirc}}{V_{max} - V_{recirc}} = \exp\left(-0.88 \times \left(\frac{y - y_{V_{max}}}{y_{0.5}}\right)^2\right) \qquad z_{V_{max}} < z \qquad (4.23b)$$

where $y_{0.5}$ is the elevation where $v_x = (V_{max} + V_{recirc})/2$ (Fig. 4.4, Right). The longitudinal variation in maximum velocity V_{max} was observed to follow closely:

$$\frac{V_{max}}{V_1} \propto \left(\frac{x - x_1}{d_1}\right)^{-1/2} \qquad\qquad z_{V_{max}} < z \qquad (4.24)$$

where x_1 is the jump toe location, and the finding is close to wall jet results (Eq. (4.20)) (Fig. 4.5).

In a bubbly air-water flow, the boundary conditions become:

$$\int_{y=0}^{Y_{90}} (1-C) \times v_x \times dy = q \qquad (4.25a)$$

$$\int_{y=0}^{Y_{90}} \rho \times (1-C) \times v_x \times \frac{\partial v_x}{\partial x} \times dy = -\frac{\partial}{\partial x}\left(\int_{y=0}^{Y_{90}} P \times dy\right) \qquad (4.25b)$$

where C is the void fraction and Y_{90} is the characteristic elevation where $C = 0.90$.

4.2.3   Advective diffusion of air bubbles

In hydraulic jumps with marked roller, air is entrapped at the jump toe that is a discontinuity between the impinging flow and the roller (Figs. 4.1 & 4.2A). The air bubble entrainment is localised at the impingement perimeter which is a line source of air bubbles, as well as a line source of vorticity. Neglecting the

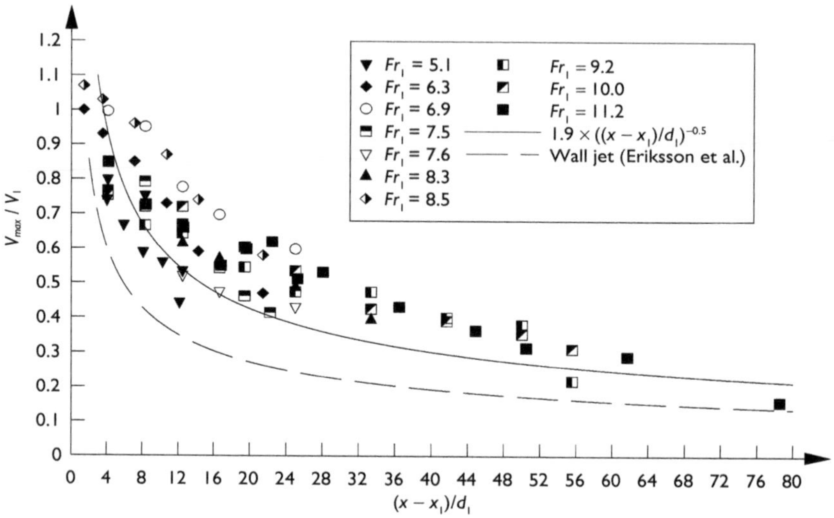

Figure 4.5 Longitudinal variation in maximum velocity in hydraulic jumps (Data: Chanson and Brattberg 2000, Kucukali and Chanson 2008, Murzyn and Chanson 2009b, Chanson 2010) – Comparison with wall jet data (Eriksson et al. 1998).

buoyancy and compressibility effects, the differential form of continuity equation for air is:

$$V_1 \times \frac{\partial C}{\partial x} + u_r \times \frac{\partial C}{\partial y} = D_t \times \frac{\partial^2 C}{\partial y^2} \tag{4.26}$$

where C is the void fraction, u_r is the bubble rise velocity assumed to be constant, and D_t is the air bubble diffusion coefficient. The development assumes implicitly a constant diffusion coefficient independent of longitudinal location and vertical elevation. Let us change the variables with $X = x + u_r/V_1 \times y$, and Equation (4.23) becomes a classical diffusion equation (Crank 1956):

$$V_1 \times \frac{\partial C}{\partial X} = D_t \times \frac{\partial^2 C}{\partial y^2} \tag{4.27}$$

The jump toe $(x = 0, y = d_1)$ acts as a point source in the $x - y$ plane. The source strength q_{air} is the entrained air flow rate per unit width. Assuming a diffusion coefficient independent of vertical elevation and an advection velocity independent of longitudinal location, Equation (4.24) may be solved analytically (Chanson 2010):

$$C = \frac{\dfrac{q_{air}}{q}}{\sqrt{4 \times \pi \times D^{\#} \times X'}} \times \left(\exp\left(-\frac{(y' - 1)^2}{\dfrac{X'}{4 \times D^{\#}}} \right) + \exp\left(-\frac{(y' + 1)^2}{\dfrac{X'}{4 \times D^{\#}}} \right) \right) \tag{4.28}$$

where $X' = X/d_1$, $y' = y/d_1$, and $D^\#$ is a dimensionless diffusion coefficient: $D^\# = D_t/(V_1 \times d_1)$. Equation (4.25) is restricted to the air-water shear layer and it is compared with experimental data in Figure 4.6.

A simpler solution was proposed earlier in the form of:

$$C = C_{max} \times \exp\left(-\frac{1}{4 \times D^\#} \times \frac{\left(\dfrac{y - Y_{C_{max}}}{d_1}\right)^2}{\dfrac{x}{d_1}}\right) \tag{4.29}$$

where $Y_{C_{max}}$ is the location where the void fraction is maximum in the developing shear layer.

4.2.4 Dimensional analysis

Any fundamental analysis of hydraulic jumps is based upon a large number of relevant equations to describe the air-water turbulent flow motion. Physical modelling may provide some insights into the flow motion subject to the selection of a suitable dynamic similarity (Liggett 1994). Considering a hydraulic jump in a smooth horizontal rectangular channel, a limited dimensional analysis gives:

$$C, \frac{V}{V_1}, \frac{v'}{V_1}, \frac{F \times d_1}{V_1}, \frac{D_{ab}}{d_1}, \frac{L_t}{d_1}, \dots$$
$$= F_1\left(\frac{x - x_1}{d_1}, \frac{y}{d_1}, \frac{z}{d_1}, \frac{V_1}{\sqrt{g \times d_1}}, \rho \times \frac{V_1 \times d_1}{\mu}, \frac{\rho \times V_1^2 \times d_1}{\sigma}, \frac{x_1}{d_1}, \frac{W}{d_1}, \frac{v_1'}{V_1}, \dots\right) \tag{4.30}$$

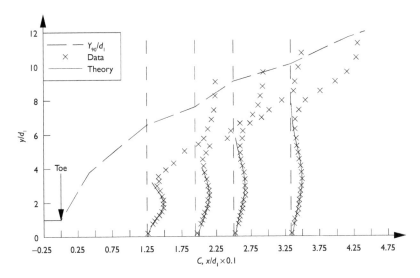

Figure 4.6 Dimensionless distribution of void fraction C (Horizontal axis: $x/d_1 \times 0.1 + C$) – Data: $Fr_1 = 9.2$, $d_1 = 0.018$ m (Chanson, 2010).

where C is the void fraction, V the air-water velocity, v' a characteristic turbulent velocity, F the bubble count rate defined as the number of bubbles detected per second in a small control volume, D_{ab} a characteristic bubble size, L_t a turbulent length scale, x the longitudinal coordinate, y the vertical elevation above the invert, z the transverse coordinate measured from the channel centreline, ρ and μ the water density and dynamic viscosity respectively, σ the surface tension between air and water, x_1 the longitudinal coordinate of the jump toe, W the channel width, v_1' a characteristic turbulent velocity at the inflow (Fig. 4.2B). Equation (4.30) expresses the turbulent flow properties at a position (x,y,z) within the hydraulic jump as functions of the inflow properties, fluid properties and channel geometry using the upstream flow depth d_1 as the characteristic length scale. In the right hand side of Equation (4.30), the 4th, 5th and 6th terms are respectively the upstream Froude number Fr_1, the Reynolds number Re and the Weber number We.

In a hydraulic jump, the momentum considerations demonstrate the significance of the inflow Froude number (Bélanger 1841, Lighthill 1978) and the selection of the Froude similitude derives implicitly from basic theoretical considerations (Liggett 1994, Chanson 2012). The Froude dynamic similarity is commonly applied in the hydraulic literature (Henderson 1966, Novak and Cabelka 1981, Chanson 2004), although the Reynolds number is another relevant dimensionless number because the hydraulic jump is a turbulent shear flow (Rouse et al. 1959, Rajaratnam 1965, Hoyt and Sellin 1989). The Π-Buckingham theorem implies that any dimensionless number may be replaced by a combination of itself and other dimensionless numbers. Thus the Froude, Reynolds or Weber number may be replaced by the Morton number Mo since:

$$Mo = \frac{g \times \mu^4}{\rho \times \sigma^3} = \frac{We^3}{Fr^2 \times Re^4} \tag{4.31}$$

When the same fluids (air and water) are used in models and prototype as in the present study, the Morton number Mo becomes an invariant and this adds an additional constraint upon the dimensional analysis. Equation (4.30) gives as simplified expression:

$$C, \frac{V}{V_1}, \frac{v'}{V_1}, \frac{F\,d_1}{V_1}, \frac{D_{ab}}{d_1}, \frac{L_t}{d_1}, \ldots = F_2\left(\frac{x-x_1}{d_1}, \frac{y}{d_1}, \frac{z}{d_1}, Fr_1, Re, \frac{x_1}{d_1}, \frac{W}{d_1}, \frac{v_1'}{V_1}, \frac{\delta}{d_1}, \ldots \right) \tag{4.32}$$

Physically it is impossible to fulfil simultaneously the Froude and Reynolds similarity requirements, unless working at full scale. It is acknowledged the air bubble entrainment is adversely affected by significant scale effects in small size models (Rao and Kobus 1971, Chanson 1997). Usually the Reynolds number was selected instead of the Weber number because prototype hydraulic jumps operate with Reynolds numbers from 10^6 to in excess of 10^9 (Fig. 4.1). For such large Reynolds numbers, the surface tension is considered of lesser significance compared to the viscous effects in the turbulent shear regions (Wood 1991, Chanson 1997, Ervine 1998). Note that the Froude and Morton similarities imply that $We \propto Re^{4/3}$ (Eq. (4.31)).

4.3   FLOW FIELD

4.3.1   Free-surface properties

Both theoretical and dimensional considerations highlighted the significance of the upstream Froude number Fr_1. For Fr_1 slightly larger than unity, the jump takes the shape of a series of stationary free-surface undulations (Henderson 1966, Chanson and Montes 1995, Chanson 2009a). For upstream Froude numbers larger than 2 to 3, the jump is characterised by a marked roller (Figs. 4.1 and 4.2). The flow properties upstream and downstream of the jump must satisfy the equations of conservation of mass and momentum (see above). This is shown in Figure 4.7 where the Bélanger equation (Eq. (4.11)) is compared with some physical data in smooth rectangular channels. The results showed a reasonable agreement despite neglecting the flow resistance.

The free-surface profile data of turbulent hydraulic jumps with marked roller shows a self-similar longitudinal profile:

$$\frac{d-d_1}{d_2-d_1}=\left(\frac{x}{L_r}\right)^{0.44} \qquad 2.4 < Fr_1 < 8.5 \qquad (4.33)$$

where L_r is the roller length. Equation (4.33) is presented in Figure 4.8 and compared with experimental data obtained with Froude numbers between 2.4 and 8.5. Note that all these data were recorded with a non-intrusive measurement technique.

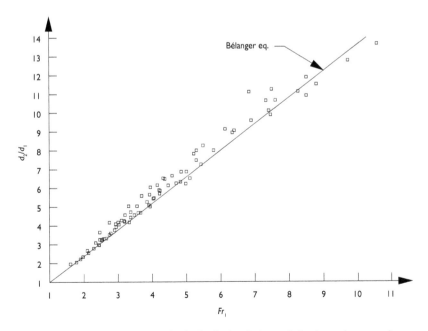

Figure 4.7 Ratio of conjugate depths for hydraulic jumps in horizontal rectangular channels – Comparison between the Bélanger equation (Eq. (4.11)) and experimental data.

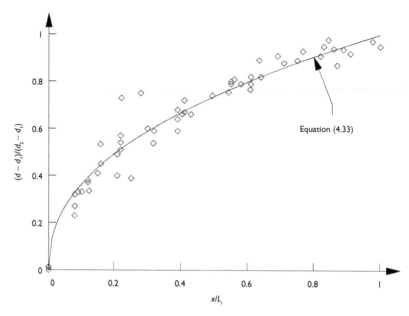

Figure 4.8 Self-similar free-surface profiles of turbulent hydraulic jumps for $2.4 < Fr_1 < 8.5$ (Data: Murzyn and Chanson 2009a, Chachereau and Chanson 2011a) – Comparison with Equation (4.33).

The results were close to theoretical solutions (Valiani 1997, Richard and Gavrilyuk 2013). The longitudinal distributions of free-surface fluctuation d' showed a marked increase in free surface fluctuations immediately downstream of the impingement point and a local maximum in free-surface turbulent fluctuations was observed in the first half of the roller (Mouaze et al. 2005, Murzyn and Chanson 2009a, Chachereau and Chanson 2011a). The free surface fluctuations increased with increasing Froude number: a monotonic increase in maximum free-surface fluctuations was observed with increasing Froude number as shown in Figure 4.9.

Physical observations highlighted that the longitudinal position of the jump toe fluctuates with time. The jump toe pulsations are believed to be caused by the growth, advection, and pairing of large-scale vortices in the developing shear layer of the jump (Long et al. 1991, Habib et al. 1994). The dimensionless jump toe frequency $F_{toe} \times d_1/V_1$ ranges between 0.003 and 0.006 independently of the upstream Froude number. For comparison, the characteristic frequency of the free-surface fluctuations tend to be larger at small Froude numbers (Chachereau and Chanson 2011a, Zhang et al. 2013).

4.3.2 Velocity field and air entrainment

The detailed turbulent flow measurements in hydraulic jumps highlight two distinct air-water flow regions (Fig. 4.2A). These are (a) the air-water shear layer and (b) the upper free-surface region. The developing shear layer is characterised by some strong interactions between the entrained air and turbulent coherent structures, associated

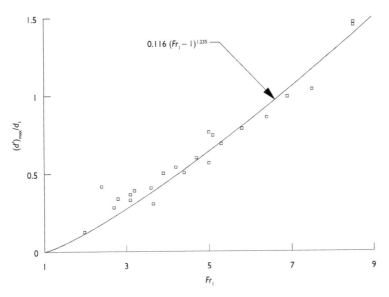

Figure 4.9 Maximum of free-surface turbulent fluctuations in hydraulic jump rollers as a function of upstream Froude number – Data: Mouaze et al. (2005), Kucukali and Chanson (2008), Murzyn and Chanson (2009a), Chachereau and Chanson (2011a).

with a local maximum in void fraction and in bubble count rate. In the shear layer, the distributions of void fractions follow an analytical solution of the advective diffusion equation for air bubbles (Eq. (4.28)) as shown in Figure 4.6. In the upper free-surface region above, the void fraction increases monotically with increasing distance from the bed from a local minimum up to unity. Figure 4.6 presents some typical vertical distributions of void fraction.

The air-water interfacial velocity distributions in the shear zone exhibit a self-similar profile that is close to that of wall jet flows (Eq. (4.23)). In the recirculation region above the shear layer, the time-averaged recirculation velocity is negative, and the re-analysis of data sets gives on average: $V_{recirc}/V_{max} \sim -0.4$ to -0.6 (Chanson 2010, Zhang et al. 2013) (Fig. 4.10). Typical data are presented in Figure 4.10. The distributions of air-water integral turbulent length scales showed similarly a monotonic increase with increasing distance from the invert (Fig. 4.11). The integral turbulent length scale characterised the size of large turbulent eddies convecting the bubbles in the hydraulic jump roller (Chanson 2007). Typical results are presented in Figure 4.11 and they highlight the relationship between integral turbulent length scale and inflow depth, as well as an increasing dimensionless length scale with increasing distance from the invert.

4.3.3   Discussion: Physical and numerical modelling

A complete description of the turbulence dynamics in strong free hydraulic jumps ($Fr_1 > 4.5$) based on prototype, experimental analysis or numerical methods is not

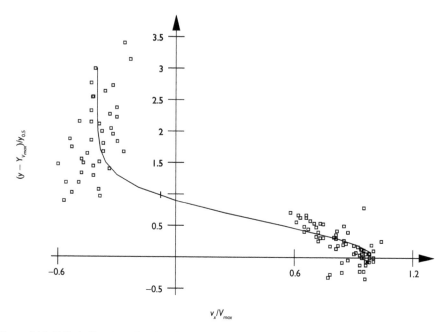

Figure 4.10 Self-similar interfacial velocity distributions in hydraulic jumps – Comparison between Equation (4.23b) and data ($Fr_1 = 10, d_1 = 0.018$ m, Chanson 2010).

yet available. Prototype measurements present some challenging risks, extremely dangerous for both equipment and operator. Laboratory studies are potentially subject to scale effects (see above). Detailed laboratory measurements of the flow properties inside strong hydraulic jumps are difficult, because the flow is highly variable in space and time, while the flow aeration restricts the use of laser light-based equipments. The numerical modelling of such highly turbulent flows also encounters considerable difficulties, particularly the accurate calculations of vortex shedding, the instantaneous position and configuration of the free surface, and interactions between turbulence and entrained air.

To date, a number of characteristics of the internal flow phenomena were measured using different techniques associated with careful data analyses, as well as some simulations in simple canonical turbulent flows. The development of accurate measurement devices has been of crucial importance for improving internal flow field knowledge in hydraulic jumps. Both experimental and numerical tools must be enhanced and complemented by prototype measurements. Numerical approaches based on continuous media of the mass and momentum conservation principles may provide information on the velocity and pressure fields leading to a better understand of energy dissipation and air-entrainment processes.

Rouse et al. (1958) presented the first detailed account of measured flow turbulence in a simulated hydraulic jump constructed in an air duct. In particular, they presented spatial distributions of turbulent quantities, including Reynolds stresses. They also used the vertically integrated momentum and energy equations

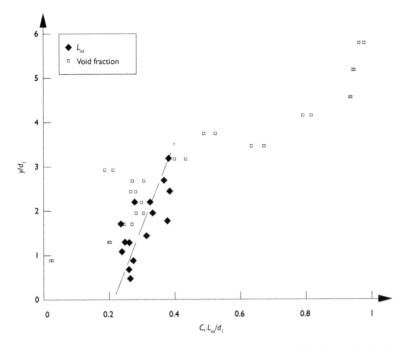

Figure 4.11 Vertical distributions of integral air-water turbulent length scales and void fraction in hydraulic jump – Data: Chanson (2007), $Fr_1 = 7.9$, $d_1 = 0.0245$ m.

to interpret the experimental results and to describe the energy and momentum budgets. Using hot film anemometry, Resch and Leutheusser (1972) investigated the distributions of Reynolds stresses and turbulence intensity in undeveloped and developed hydraulic jumps for Froude numbers $Fr_1 = 2.85$ and 6. Further studies characterised hydraulic jump features, including Rajaratnam (1965,1967) and Mccorquodale and Khalifa (1983). More recently, higher accuracy, quality data were obtained using dual-tip phase-detection probes, allowing significant improvements on the characterisation of the air-water flow field inside hydraulic jump (Chanson and Brattberg 2000, Murzyn et al. 2005, Chanson 2009a). Research works focusing on flow visualization techniques also emerged, allowing new qualitative and quantitative information: e.g., image processing procedures (Mossa and Tolve 1998, Leandro et al. 2012), BIV technique (Ryu et al. 2005). However such visualisations techniques are limited to two-dimensional flows in the vicinity of sidewalls and need specific calibration. New techniques based upon films and photographs taken by boroscope are promising but not yet completely developed.

The hydraulic jump turbulence dynamics in such high Reynolds number flows can be studied numerically using mass and momentum conservation equations in their three-dimensional form, Navier-Stokes and ensemble-averaged Reynolds-average Navier-Stokes equations coupled with the Volume Of Fluid (VOF) method (Hirt and Nichols 1981) to represent arbitrary free surfaces: e.g., Lemos (1992) for $Fr_1 = 2.8$,

Qingchao and Drewes (1994) for Fr_1 = 4–7, Carvalho (2002) and Carvalho et al. (2008) for Fr_1 = 6. Fractional Area-Volume Obstacle Representation (FAVOR) algorithm (Hirt and Sicilian, 1985) may be used to describe details of the internal obstacles and other geometry particularities. The Navier-Stokes equations and the equation governing the VOF-function $F = F(x,y,z,t)$, whose value is 1 for a point occupied by the fluid and zero elsewhere, are:

$$\nabla \cdot \left(\theta \vec{v}\right) = 0 \tag{4.34}$$

$$\frac{\partial v_i}{\partial t} + \frac{1}{\theta}\nabla\left(\theta \vec{v} \otimes \vec{v}\right) = \vec{g} - \frac{1}{\rho}\nabla p + \frac{1}{\theta}\nabla\left(\theta \nu \nabla \vec{v}\right) + s_i \tag{4.35}$$

$$\frac{\partial F}{\partial t} + \frac{1}{\theta}\nabla\left(\theta \vec{v}\right) = 0 \tag{4.36}$$

where $\vec{v} = (v_x(x, y, z, t), v_y(x, y, z, t), v_z(x, y, z, t))$ is the velocity vector, $p(x,y,z,t)$ is the pressure, $v = \mu/\rho$ is the kinematic viscosity, $\vec{g} = (g_x, g_y, g_z)$ is the body force term, s_i is a term that takes into account other influences to the velocity field such as turbulence, air, sediment flow interactions or boundary influence and $\theta(x,y,z,t)$ is the porosity function whose value is 0 for a point inside an obstacle and 1 for a point that can be occupied by the fluid. If $\theta = 1$ for all points of the domain, Equations (4.34) and (4.35) reduce to the Navier-Stokes equations.

Direct Numerical Simulation (DNS) may model properly the hydraulic jump turbulence dynamics if the mesh could describe both large and small viscous scales. This would require a number of mesh points proportional to $Re^{9/4}$ and is not yet possible. Either details of large vortices and free-surface rolling configurations are priority and energy dissipation is not properly represented (e.g. Figs. 4.12 & 4.13 Left), or turbulence models, such as those based on Reynolds-averaged or renormalization group, are applied and the energy dissipation could be approximately represented but with a loss of details in terms of fluid dynamics (Fig. 4.13 Right).

Two families of turbulence models are those based upon Reynolds-Averaged or Renormalization Group (RNG) characterising turbulence by a specific turbulent

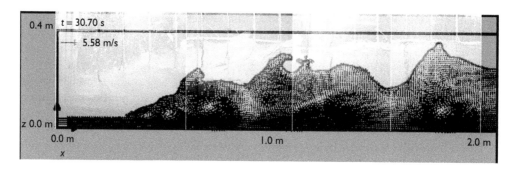

Figure 4.12 Computed velocity field and free surface configuration for the whole computational domain superimposed onto a snapshot of the hydraulic jump in the laboratory experiments (Fr_1 = 6).

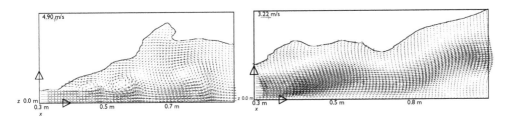

Figure 4.13 Detail aspects of the vortices and free surface configuration for the numerical simulation of hydraulic jump ($Fr_1 = 6$): (left) without including any type of turbulence model, considering the coefficient of kinematic viscosity of $v = 1.15 \times 106$ m² s⁻¹ and a very small time increment Δt; (right) using the RNG k–ε turbulence model. (Data Carvalho et al. 2008).

kinetic energy $k(x, y, z, t)$ and its rate of dissipation $\varepsilon(x, y, z, t)$ per unit mass. These models required two closure equations for k and ε:

$$\frac{\partial k}{\partial t} + \frac{1}{\theta}\nabla\left(\theta \vec{v} k\right) = \frac{1}{\theta}\nabla\left[\theta\left(v + \frac{v_t}{\sigma_k}\right)\nabla k\right] + P - \varepsilon \tag{4.37}$$

$$\frac{\partial \varepsilon}{\partial t} + \frac{1}{\theta}\nabla\left(\theta \vec{v} \varepsilon\right) = \frac{1}{\theta}\nabla\left[\theta\left(v + \frac{v_t}{\sigma_k}\right)\nabla \varepsilon\right] + C_{\varepsilon1}\frac{\varepsilon}{k}P - C_{\varepsilon2}\frac{\varepsilon^2}{k} - R \tag{4.38}$$

where P is the shear production term, R is the rate-of-strain term given by:

$$R = \frac{C_\mu \eta^3\left(1 - \eta/\eta_0\right)}{1 + \beta\eta^3}\frac{\varepsilon^2}{k}$$

and the constants are $\sigma_k = 1.0$, $\sigma_\varepsilon = 1.3$, $C_{\varepsilon1} = 1.42$, $C_{\varepsilon2} = 1.68$, $C_\mu = 0.085$, $\eta_0 = 4.38$ and $\beta = 0.012$; the eddy-viscosity is defined in terms of k and ε by the expression $v_T = C_{\varepsilon2}\varepsilon^2/k$; $\eta = Sk/\varepsilon$; $S = 2S_{ij}S_{ij}$ and $S_{ij} = 0.5(\partial v_i/\partial x_j + \partial v_j/\partial x_i)$.

The Large Eddy Simulation (LES) formulation is based on the application of spatial filters in the Navier-Stokes equations to model directly the large-scales whereas small-scales (Sub-Grid Scales) are approximated, thus requiring less computational resources than DNS. The LES models preserve more turbulence properties allowing an estimate of force fluctuations associated with turbulence motion. The governing equations are thus transformed and the solution is a filtered velocity field:

$$\frac{\partial \bar{v}_i}{\partial t} + \frac{1}{\theta}\nabla\left(\theta \vec{\bar{v}} \otimes \vec{\bar{v}}\right) = \vec{g} - \frac{1}{\rho}\nabla\bar{p} + \frac{1}{\theta}\nabla\left(\theta v \nabla\vec{\bar{v}} + T_{ij}\right) \tag{4.39}$$

where $T_{ij} = \left(\overline{\bar{v}_i \bar{v}_j} - \overline{\bar{v}_i}\,\overline{\bar{v}_j}\right) - \left(\overline{\bar{v}_i v_j'} + \overline{v_i' \bar{v}_j}\right) - \overline{v_i' v_j'}$ is the "*subgrid-scale tensor*" given by the addition of three tensors: the Leonard tensor (first two terms) representing the interactions among large scales, the Clark tensor (third and fourth terms) for cross-scale interactions between large and small scales and the Reynolds stress-like term characterising interactions among the Sub-Grid Scales (SGS) (Aldama, 1990).

Simpler SGS models lumped the effects of turbulence into a turbulent viscosity modelled proportionally to a length scale times a measure of velocity fluctuations at that scale

$$\nu_T = (cL)^2 \sqrt{2S_{ij}S_{ij}} \qquad (4.40)$$

where c is a constant having a typical value in the range of 0.1 to 0.2 and the length scale $L = (dxdydz)^{1/3}$. The question of how fine a grid resolution is needed to resolve a filtered velocity field is discussed in Pope (2000, Chapter 13).

Enhancements in turbulence models and sub-grid scales, modelling free-surface tension were recently developed (Liovic and Lakehal, 2012) and they might be important to model air entrainment. Such models are becoming more and more viable tools, because of the increase in computational speed, memory and parallel processing as well as open source numerical models that have been made available.

Disregarding air entrainment in numerical models can produce errors ranging from 25% to 100% for the turbulence in spilling breaker zones (Lubin et al. 2011). To describe mathematically the air entrainment and the two phase dynamics interacting, models have to use the Navier-Stokes equations in their multiphase form, including conservation equations for both phases, water and air, resulting from ensemble average operations who assume bubble-bubble distance between intermediate to large scales (Drew and Passman 1999, Bombardelli 2012). New terms appear in the equations, including the possibility of mass transfer between phases in the mass conservation equation, a new shear stress term coming from the advective-acceleration terms and averaging procedure, inter-phase forces in momentum equations that leads to drag, lift, virtual mass and Magnus force, which require sub-models as well as interfacial/ wall transfer closure laws for the air bubbles and the water.

The air entrainment occurs when turbulence overcomes the stabilizing forces of gravity and surface tension and eddies can rise above a free surface, trap air and carry it back into the body of the water. Turbulence and destabilising effects of the free-surface are then the main causes of air ingestion. Special care should be given to the non-linear convective terms and the viscous term with variable viscosity and surface tension description in Navier-Stokes equations. Hirt (2003) evaluated the energy per unit of volume associated with the disturbance and the energy of the stabilizing forces effect by gravity and surface tension and if the first is larger than the second, a volume of air is allowed to enter the flow. Ma et al. (2011) used a parametric equation to describe air entrainment occurring when the air cavities just below the interface are drawn into the liquid at a rate that is faster than the downward motion of the local air/liquid interface. They developed a comprehensive and accurate subgrid air entrainment model and used it to model air entrainment and transport in a hydraulic jump. This process is not well described with traditional Reynolds average models that are unable to treat properly these scales and then unable to simulate air entrainment at the wavy surface, as it is with LES.

Despite the difficulties of numerical modelling, the results give insights into the internal structure of the hydraulic jump as well as hydrodynamic pressures and forces on the bottom, chute blocks, baffles piers and end sills that are important to understand the internal processes (Ferziger and Perić 1996, Carvalho 2002, Carvalho et al. 2008).

4.4 HYDRAULIC DESIGN OF STILLING BASINS

4.4.1 Presentation

A stilling basin is a concrete structure that contains and promotes turbulent kinetic energy dissipation to decrease the high velocity flow erosive power (Fig. 4.14). Thus, it protects the stream-bed by dissipating energy and preventing damage/scouring around the structure. Hydraulic jumps are often used to dissipate the kinetic energy of high velocity water flows downstream of a steep chute spillway, outlet works, culvert structures and other hydraulic structures. Hydraulic jump stilling basins are relatively expensive structures, used when the geotechnical characteristics are not good enough, e.g. with soft and fractured rocks or mudslides.

Hydraulic design guidelines provide stilling basin geometric characteristics to contain a specific hydraulic jump in the stilling basin and tail water control to guarantee its occurrence within the basin. Geometric dimensions and features depend upon the inflow water depth and velocity (d_1 and V_1), hence the Froude number, and the required energy dissipation rate. Because of the differences in hydraulic jump characteristics, there are two main types of hydraulic jump stilling basins: low Froude number ($2.5 < Fr_1 < 4.5$) and high Froude number ($Fr_1 > 4.5$), corresponding to low- and high-head basins respectively. The most effective hydraulic jumps are in the range of $Fr_1 = 4.5$ to 10, resulting in a comparatively shorter basin.

Guidelines give dimensions for geometric basin details and predict the free-surface profile along the stilling basin, indicating the required height of basin sidewalls, reinforcement of basin floor, downstream water depth, as well as the roller length, stilling basin length, hydrodynamic pressures on the floor and eventually on basin appurtenances. The design may include expansion, drops, steps, blocks, baffles and sills, typically used to decrease the basin length and stabilise the jump toe position (Fig. 4.14). An expansion and/or depression is usually considered to increase the Froude number, by converting potential energy into kinetic energy, thus allowing the flow to expand, drop, or both. Blocks, baffles and sills are used to minimise the length of hydraulic jump. The devices interfere with the jump and generate high levels of turbulence and energy dissipation, although they increase the risks of cavitation. Their application is limited to a range of velocity, unless some aeration devices are included to prevent cavitation damage. The addition of blocks and end sills do not result in significantly higher maximum negative and positive pressures than those for basins without blocks and sills (Toso and Bowers, 1988). Increasing the boundary roughness and injection of additional discharge has not received much attention because of associated cavitation and stability issues. The end sill has another important function, namely to reduce scouring at the end of the basin (Lemos 1983) (Fig. 4.15).

A number of standard basin designs were developed. Some energy dissipators were derived with sudden expansions, abrupt deflections, baffle basins, drop structures, counterflows, rough walls, vortex devices, and spray inducing devices. Special care must be considered to follow the recommended criteria for stilling basins. When the

Figure 4.14 Photographs of hydraulic jump stilling basins (A) Hinze dam spillway stage 3 (Australia) on 25 April 2013 (Courtesy of Davide Wüthrich): the stilling basin was designed based upon a physical study and each block is 3.2 m high. (B) Stilling basin in Jeju Island (Korea): on 26 May 2013 at low tide and on 28 May 2013 after a storm. (C) Loundoun weir stilling basin (Australia) on 16 January 2009 (Courtesy of Damien Roman).

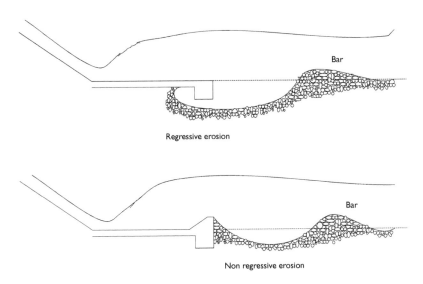

Regressive erosion

Non regressive erosion

Figure 4.15 Detail aspect of the end sill influence (adapted from Lemos, 1983).

velocity or flow rate are outside the recommended range, it is critical to perform some physical modelling of individual structures. In the great majority of cases, the costs of laboratory experiments is negligible compared to the benefits of an optimum design. Examples of well-known issues include: if the baffle blocks are far too upstream, wave action in the basin may develop; if too far downstream, a longer basin will be required; if too high, wave motion can be produced; if too low, jump may sweep out (FHWA, 2006).

The design of a stilling basin structure as shown in Figure 4.16 involves:

–   Determination of the inflow velocity V_1 and depth d_1 entering the basin, and evaluation of Froude number Fr_1 (Eq. (4.8) or (4.14)).
–   Investigation of the river cross-section, definition of the downstream channel dimensions, and determination of the tailwater depth TW.
–   Estimation of the conjugate depth for the outlet conditions using Equation (4.4) or (4.8); check energy equation to verify if a basin is needed – compare the conjugate depth d_2 and tailwater depth TW; if $d_2 < TW$ for the entire range of discharges, there is sufficient tailwater and a jump will form without a basin although the floor might need to be reinforced; if $d_2 = TW$, this is an ideal case; if $d_2 > TW$ for the entire range of discharges, it is recommended to use apron protection, baffles and sills to create hydraulic jump within the basin; if $d_2 > TW$ for lower discharges and $d_2 < TW$ for higher discharges, it is recommended a stilling basin and sloping apron; the opposite indicates a stilling pool.
–   Selection of potential standard stilling basins based upon the inflow Froude number and discharge (see next section).
–   Determination of invert elevation, basin length, basin elements (bottom elevation h_1, basin width W_B, steps appurtenances dimensions and slopes) and

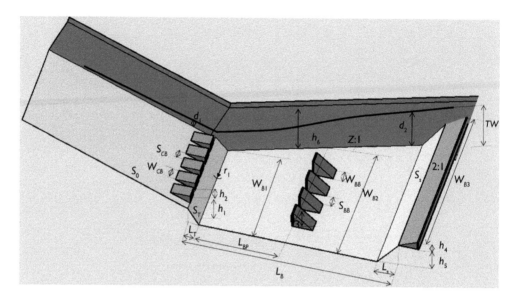

Figure 4.16 Definition sketch of a hydraulic jump stilling basin nomenclature.

verification that sufficient tailwater level exists to force the hydraulic jump inside the basin.
– Determination of the radius of curvature r_1 for the slope changes entering the basin.
– Finally an analysis of hydraulic jump in the stilling basin, their external and internal characteristics including average pressure distributions and hydrodynamic pressures and forces and their extreme probability values.

4.4.2   Standard stilling basins

Several standard stilling basin designs were developed in the 1940s to 1960s. These basins were tested in models and prototypes over a considerable range of operating flow conditions. Prototype performances are well understood, and these designs can be selected without further model studies within their design specifications.

A number of standard stilling basin design procedures are available from several institutes which conducted extensive experimental investigations. Among these, the US Bureau of Reclamation (USBR or BUREC) proposed five types and later ten types of USBR basins (Peterka 1978, Hager 1992). Some stand out, in particular types II and III for high Froude numbers and type IV for low Froude numbers. The Saint Anthony Falls (SAF) laboratory, Public Works Department (PWD), Waterways Experimental Station (WES), R.S. Varshney and Indian Standard Stilling Basin are also often referenced for low Froude number basins ($Fr_1 < 4.5$) and downstream culverts stilling basins. FHWA (2006) discussed some designs. VNIIG in Leningrad developed four types of stilling basins for inflow Froude numbers between 2.5 and 10 (Yuditskii, 1963).

4.4.2.1   Hydraulic jump stilling basin USBR/BUREC

The USBR stilling basins were first presented by Bradley and Peterka (1957a,b,c,d,e,f) (also Peterka 1978, BUREC 1980). Their designs were based upon experimental testing in stilling basins and confirmed with successful prototype operations. Figure 4.17 shows the relationship between the dimensionless basin length L_B/d_2 and tailwater depth TW/d_1, and the inflow Froude number Fr_1 for the USBR stilling basin types I, II and III, to ensure that the hydraulic jump is constrained within the basin. The following correlations could be used instead of the graph.

Type I: $L_B/d_2 = -5\text{E-}05Fr_1^3 - 0.0036\ Fr_1^2 + 0.0816\ Fr_1 + 5.6544$ with $R^2 = 0.9927$
Type II: $L_B/d_2 == 0.0009\ Fr_1^3 - 0.0385\ Fr_1^2 + 0.5328\ Fr_1 + 1.9551$ with $R^2 = 0.9913$
Type III: $L_B/d_2 == 0.0005\ Fr_1^3 - 0.0222\ Fr_1^2 + 0.3271\ Fr_1 + 1.1558$ with $R^2 = 0.9897$
Type IV: $L_B/d_2 = 0.0411\ Fr_1^3 - 0.5838\ Fr_1^2 + 3.0198\ Fr_1 + 0.3164$ with $R^2 = 0.9982$

4.4.2.1.1   USBR Type I ($1.7 < Fr_1 < 2.5$)

The USBR Type I is a rectangular stilling basin with a horizontal bottom, no appurtenances, no chutes blocks, no baffles or sills, which contains a classical hydraulic jump. Usually it is not recommended because of costs and safety, as the basin length is large and the hydraulic jump is sensitive to downstream level variation.

4.4.2.1.2   USBR Type II ($Fr_1 > 4.5$; $V_1 > 18$ m/s)

The USBR type II stilling basin (Fig. 4.18) was developed for high dams, earth dam spillways and large canal structures. The design includes chute blocks and dentated end

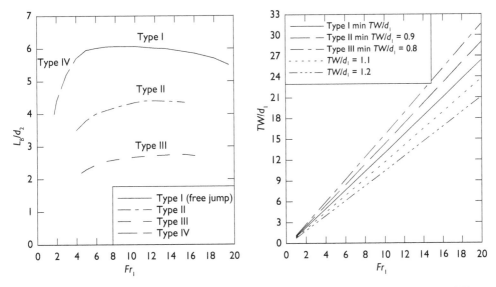

Figure 4.17 Hydraulic jump stilling basin length and tail water depth for USBR Type I, II and III (adapted from Peterka, 1978).

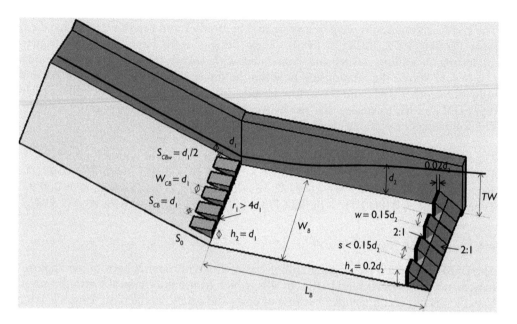

Figure 4.18 Hydraulic jump stilling basin USBR Type II.

sill, thus reducing by 30% the required length compared to a classical hydraulic jump. Considering the high inflow velocities, baffles were not included, and a tailwater depth 5% higher than the conjugate depth is recommended. In this type, the end sill has a stabilizing effect and also a dissipative reason. The procedure to design USBR type II consists first in calculating the required length (Fig. 4.17, Left) and the conjugate water depth to set apron elevation which is (with a 5% factor of safety): $TW = 0.475 \times d_1 \times ((1 + 8 \times Fr_1^2)^{0.5} - 1)$ (Fig. 4.17, Right). The number of recommended chute blocks is $n_{cb} = W_B/(2 \times d_1)$ and their height, width and spacing should be about d_1 except at the wall where should be around $d_1/2$ wide to reduce spray ($w_{cb} = s_{cb} = W_B/(2 \times n_{cb})$). Although it is important to maintain spacing equal, it could be reduced specially for narrow basins. They could be incorporated in upstream surface which slope could vary between 0.6:1 and 2:1. Such an upstream slope does not have significant influence on the basin behaviour. Attention should be given to the intersection between chute and basin apron that should be formed by a curve with radius larger than $4 \times d_1$, especially if upstream slope is higher than 1:1. The height of the end sill should be equal to $0.2 \times d_2$ and the maximum spacing should be less than $0.15 \times d_2$. Additional hydraulic studies are needed for heads or flows higher than 60 m and 45 m³/s respectively (Lemos 1983).

4.4.2.1.3   USBR Type III (Fr₁ > 4.5; V₁ < 18.3 m/s; q < 18.6 m³/s/m)

The USBR Type III stilling basin (Fig. 4.19) includes chute blocks, baffles piers and end sill. It was developed for gravity dam, earth dam spillways and large canal structures to reduce by 45% the required length for a classical hydraulic jump (2/3 Type II length). With the inclusion of baffles, the inflow velocities are restricted

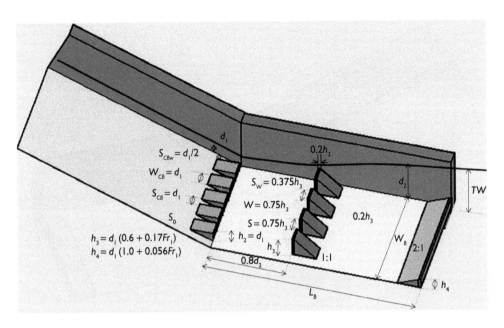

Figure 4.19 Hydraulic jump stilling basin BUREC Type III.

to avoid cavitation damage to the concrete surface and reduce the impact force to the blocks. The required length is shown in Figure 4.17 (Left). Baffle blocks should be placed at $0.8 \times d_2$. Their height (h_3) and the end sills height (h_4) are recommended as follows: $h_3 = d_1 \times (0.175 \times Fr_1 + 0.6)$; $h_4 = d_1 \times (0.0571 \times Fr_1 + 0.9714)$ (Fig. 4.19). The downstream water level should be at least $0.832 \times d_2$ to guarantee the formation of a hydraulic jump inside the basin. The basin sidewalls must be vertical, rather than trapezoidal, to insure proper performances of the hydraulic jump dissipator.

4.4.2.1.4 USBR Type IV (2.5 < Fr₁ < 4.5)

The USBR Type IV stilling basin (Fig. 4.20) was developed for oscillating hydraulic jumps. It includes chute deflector blocks to reduce the instability of the oscillating jump and a continuous end sill. The number of blocks could be estimated by $n_{cb} = W_B/ (2.625 \times d_1)$; the block width and block spacing are $w_{cb} = W_B/3.5$ (with typically $0.75 \leq w_{cb}/d_1 \leq 1$) and $s_{cb} = 2.5 \, w_{cb}$, respectively. The end sill height is $h_4 = d_1 (0.0536 \, Fr_1 + 1.04)$ and its slope should be 2 H:1 V.

4.4.2.2 Saint Anthony Falls – SAF stilling basin (1.7 < Fr₁ < 17)

The SAF hydraulic jump stilling basin (Fig. 4.21) includes chute blocks, baffle blocks and end sill, reducing the basin length compared to a free hydraulic jump. It is recommended at the downstream of small structures, such as culverts, weirs and outlet

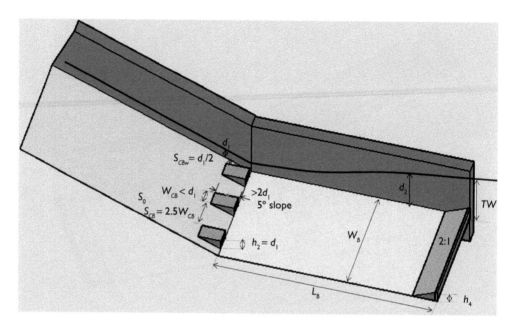

Figure 4.20 Hydraulic jump stilling basin BUREC Type IV.

works. The SAF design was refined from the first design. Nowadays the basin length should be as follows:

$$L_B = \frac{4.5d_2}{CFr_1^{0.76}}; \quad C = \begin{cases} 1.1 - \dfrac{Fr_1^2}{120} & \text{for } 1.7 < Fr_1 < 5.5 \\ 0.85 & \text{for } 5.5 < Fr_1 < 11 \\ 1 - \dfrac{Fr_1^2}{800} & \text{for } 11 < Fr_1 < 17 \end{cases} \qquad (4.41)$$

while W_B should be function of pipe diameter D (Fig. 4.21). The basin width W_B is taken as the larger of the culvert diameter and the value:

$$W_B = 1.7D_0 \frac{Q}{g^{0.5}D_0^{2.5}} \qquad (4.42)$$

or for upstream channel W_B must be equal to channel width. W_{B1} takes the value:

$$W_{B1} = \begin{cases} D & \text{for } V_1 < 6 \text{ m/s} \\ 2.5 \text{ to } 3D & \text{for } 6 < V_1 < 12 \text{ m/s} \end{cases} \qquad (4.43)$$

The number of chute blocks should be $n_{cb} = W_{B1}/(1.5 \times d_1)$ which give block width and spacing $W_{B1}/2n_{cb}$ having a half block placed at the basin wall. Baffles should not

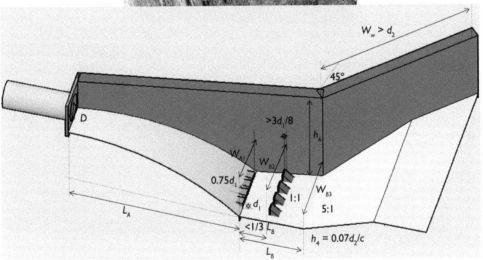

Figure 4.21 Hydraulic jump SAF stilling basin.

be placed at the wall nor in a distance lower than $3 \times d_1/8$. Its height should be equal to d_1 and the top thickness around $0.2 \times d_1$ with back slope of the block 1:1. The distance between chute blocks and baffles should be no greater than $L_B/3$. The number of baffles is $n_b = W_{B2}/(1.5 \times d_1)$; the width and spacing are $W_{B2}/(2 \times n_b)$ and, as there are n_b baffles and $n_b - 1$ spaces, the remaining width should be divided equally for calculate space between outside baffle and wall. The width at baffles should be obstructed by baffles between 40 and 55% and could be calculated by:

$$W_{B2} < W_{B1} + 2zL_B/2 \qquad (4.44)$$

The height of the end sill is $h_4 = 0.07 \times d_2/C$ and its slope should be set to 2 H:1 V. The length of the sill is

$$W_{B3} = W_B + 2zL_B \tag{4.45}$$

The height of the sidewall above the floor of the basin is given by:

$$h_6 > d_2(1 + 1/(3C)) \tag{4.46}$$

Finally the wingwalls should be equal in height and length to the stilling basin sidewalls. The top of the wingwall should have a 1 V:1 H slope. The best overall conditions are obtained if the triangular wingwalls are located at an angle of 45° to the outlet centerline.

4.4.2.3   PWD (Public Works Department) and WES (Waterways Experiment Station) stilling basins

The PWD basin (Fig. 4.22) is recommended for a stilling structure downstream of circular pipes with diameters between 450 mm and 1,850 mm and for drop lower than $3 \times D$ and velocities less than $U < 2\sqrt{gD} = 6.264 \times D^{0.5}$. The WES basin (Fig. 4.23) differs from the PWD design because it is longer and presents a smaller divergence angle. The length should be around $5 \times D$ and h_3 is calculated as:

$$\frac{Q}{D^{5/2}} \leq 2.9\frac{h_3}{D}\left(\frac{L}{D}\right)^{0.4(D/h_3)^{1/3}} \tag{4.47}$$

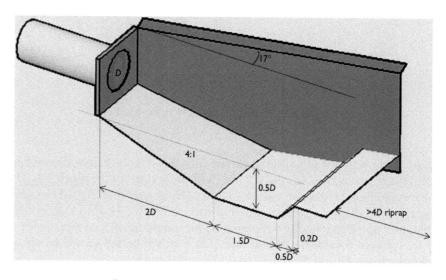

Figure 4.22 Hydraulic jump stilling basin PWD.

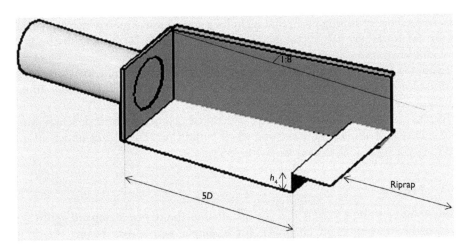

Figure 4.23 Hydraulic Jump stilling basin WES.

4.4.2.4 FWHA stilling basins

Downstream of culverts and channels, FWHA (2006) recommends a basin with three parts; first with a slope S_T and an expansion from W_0 to W_B, then a horizontal part and a final part with slope S_S, both rectangular (Fig. 4.16). The slope could be $S_S = S_T = 0.5$ (0.5 V:1 H) or 0.33 (0.33 V:1 H) and the elevations $h_1 = d_2 - TW$ and $h_5 = L_S \times S_S$. W_B should be lower than:

$$W_B \leq W_0 + \frac{2L_T \sqrt{S_T{}^2 + 1}}{3Fr_2} \qquad (4.48)$$

The lengths of the basin are calculated as follows: L_B from Figure 4.17 and L_S and L_T as:

$$L_S = \frac{L_T (S_T - S_0) - L_B S_0}{S_S + S_0}; \quad L_T = \frac{z_0 - z_1}{S_T} \qquad (4.49)$$

If the transition slope is 0.5 V:1 H or steeper, it is recommended a circular curve at the transition to the stilling basin with radius:

$$r = d_0 / e^{\frac{1.5}{Fr^2}} - 1 \qquad (4.50)$$

It is also advisable to use the same curved transition going from the transition slope to the stilling basin floor.

4.4.2.5 VNIIG Gunko, Lyapin and Kumin stilling basins

Gunko is a compact stilling basin that includes a continuous block which high depends from Froude number. It is recommended for chutes under 40 m and discharges per unit width lower than 80 m²/s.

Lyapin is a compact stilling basin which includes a row of blocks which height is function of d_1 as shown on Figure 4.24. It is recommended for chutes under 20 m and discharges per unit width lower than 80 m²/s.

Kumin is a compact structure with a detailed geometry block developed to dissipate energy and spread the incoming jet. It is indicated for chutes under 30 m and discharges per unit width lower than 100 m²/s.

4.4.3 Notes

Hydraulic jump stilling basins may be installed at the toe of a stepped spillway and downstream of roughened channels, for example with cross beams. Preliminary design criteria were proposed for stepped spillways with hydraulic jump formation using simple shapes and adequate relations of critical depth/step height (Carvalho and Martins 2009, Bung et al. 2012) (see also Chapter 4). Hydraulic jumps over cross beams were study intensively by Morris (1968) who called this tumbling flow. Recently, empirical relations to predict flow characteristics in the roughened channels with cross beams were proposed by Carvalho and Lorena (2012).

Although considerable research was dedicated to hydraulic jump stilling basins and while standard stilling basins present a generally good operational record, new stilling basin designs are still proposed. Special features, such as different forms of basins like trapezoidal section, divergent walls and corrugated aprons to dissipate additional energy, are sometimes required to be more economical as well as to deal with different tailwater situations.

4.5 PROTOTYPE OPERATION AND EXPERIENCE

A number of prototype tests on hydraulics structures were conducted prior to and shortly after World War II (Campbell and Pickett 1959). Other prototype measurements were reported by Preobrazhenskii (1958), Yuditskii (1960–64) and Ivoilov (1982) who measured hydrodynamic pressures and verified deviation of hydrodynamic forces between prototype and model about 20%. More recent data were discussed by Deng et al. (2007).

A number of prototype studies were undertaken, typically after the structures experienced damage. The investigations focused on damage by cavitation (collapse of vapor bubbles formed by pressure changes within a high-velocity water flow) and abrasion (erosion of concrete caused by water-transported sediments, ice and debris). The prototype observations indicated several cases of stilling basin erosion, mostly following unusual conditions during the construction period, poor structural shapes and insufficient operation of the bed (Hager 1992). As the result of inadequate design, construction, and operational and environmental changes, damage occurred on stilling basins steps, abrupt sluice expansions and on the baffle

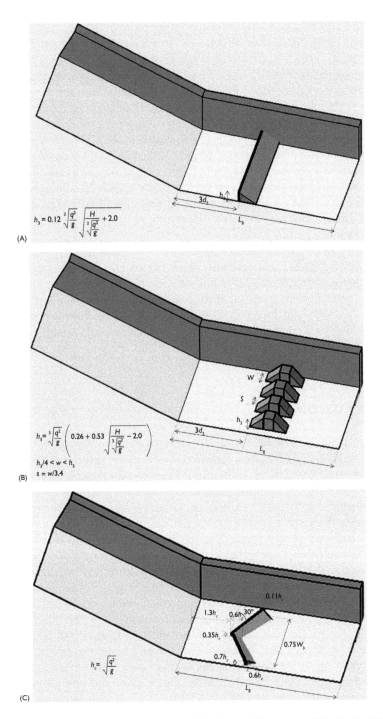

$$h_3 = 0.12 \sqrt[3]{\frac{q^2}{g}} \sqrt{\frac{H}{\sqrt[3]{\frac{q^2}{g}}} + 2.0}$$

(A)

$$h_3 = \sqrt[3]{\frac{q^2}{g}} \left(0.26 + 0.53 \sqrt{\frac{H}{\sqrt[3]{\frac{q^2}{g}}}} - 2.0 \right)$$

$h_3/4 < w < h_3$

$s = w/3.4$

(B)

$$h_c = \sqrt{\frac{q^2}{g}}$$

(C)

Figure 4.24 VNIIG stilling basins. (A) Gunko stilling basin. (B) Lyapin stilling basin. (C) Kumin stilling basin.

piers sides. Most were experienced for discharges lower than the design discharge. Separation and recirculation cavities often began to form near curves and offsets in a flow boundary, or at the centres of vortices. There were several cases in which baffle blocks connected to the basin training walls generated eddy currents downstream of baffles, resulting in localised damage. Graham et al. (1998) concluded that the magnitude and duration of operation had some important role in terms of stilling basin wall and floor slab performances. The main causes of major damage were pressure fluctuations, including extreme values. These occurrences motivated several studies of hydrodynamic pressures and forces (Toso and Bowers 1988, Pinheiro et al. 1994, Fiorotto and Rinaldo 1992, Teixeira, 2008). Teixeira (2008) studied scale effects by comparing several scale models (1:100, 1:50 1:32) and prototype data. The results showed that the average pressures in scale models are slightly larger, while no significant scale effect was observed in terms of pressure fluctuations. However scale effects cannot allow any proper study of cavitation, air water flows and scouring in physical models.

Recent prototype measurements of pressure fluctuation were performed in The Dalles Dam stilling basin equipped with baffle blocks and end sill (Deng et al. 2007). Statistical analyses of pressure sensor mounted on baffles below one of two spillways gates showed that the mean pressure increased with the water depth, being higher on the upstream side and lower in flow separation zones. Power spectra indicated an increased low-frequency spectral power with increased discharge, typical spectral decay rates (f^{-1} for lower frequencies and $f^{-5/3}$ in higher frequency), and the existence of spectral peaks of possible hydraulic origin.

4.6 CONCLUSION

The transition from a high-velocity to fluvial open channel flow is a flow singularity called a hydraulic jump. The flow properties across the jump may be solved using the continuity and momentum principles. Theoretical considerations highlight the significance of the upstream Froude number Fr_1 and show that the effects of flow resistance become small for Froude numbers greater than 2 to 3. The turbulent hydraulic jump with roller is characterised by strong air bubble entrainment, spray and splashing. A key feature of the turbulent shear region is the vertical velocity distributions which present a self-similar shape similar to that observed in wall-jets. The impingement perimeter at the jump toe acts as a source of turbulent vorticity as well as a line source of entrained air. An analytical solution of the advective diffusion equation provides a two-dimensional solution and analytical results are found to be in a good agreement with physical measurements of void fraction in the developing shear region.

Hydraulic jumps are commonly used at the toe of weirs and dam spillways to dissipate the flow energy and to convert the flow from supercritical to subcritical conditions. Most turbulent kinetic energy is dissipated in the hydraulic jump, sometimes assisted by appurtenances including step, baffles and end sill to enhance the turbulence, control the jump position and reduce scour downstream. The hydraulic design of a stilling basin must ensure a safe dissipation of the flow kinetic energy and a maximum rate of energy dissipation with minimum size (and cost) of the structure. A number of standard stilling basin designs were developed in the 1940s to 1960s (section 4). These basins were tested in laboratory models and prototypes over a considerable range of operating flow conditions. Their performances are well known, and they may be selected without further model studies, within well-specified conditions. Outside of these, some physical modelling is strongly recommended. Importantly design engineers must ensure that the hydraulic jump stilling structure will operate safely for a wide range of flow conditions, including non-design flow conditions. Damage (scouring, abrasion, cavitation) to the basin and to the downstream natural bed could occur for a number of reasons, if a thorough hydraulic design is not properly undertaken.

Despite several centuries of research, the hydraulic jump remains a fascinating flow motion that is still poorly understood. Numerical modelling of hydraulic jumps is not easy because the hydraulic jump flow encompasses many challenges including two-phase flow, turbulence, and free surface. Present numerical techniques (LES, VOF) may be applied to large-size turbulent flows comparable prototype civil and engineering applications, but they lack microscopic resolutions and are not always applicable to two-phase flows. Other techniques (DNS) provide a greater level of small-scale details but are limited to turbulent flows in relative small-size facilities. Future studies of hydraulic jumps may be based upon some 'composite' models embedding physical, theoretical and numerical studies.

ACKNOWLEDGEMENTS

The authors thank their students, former students and co-workers. The financial support of the Australian Research Council (Grants DP0878922 & DP120100481) is acknowledged.

REFERENCES

Aldama, A.A. (1990). *Filtering Techniques for Turbulent Flow Simulation.* Lecture Notes in Engineering, Springer-Verlag.

Bélanger, J.B. (1841). *Notes sur l'hydraulique.* ('Notes on hydraulic engineering') Ecole Royale des Ponts et Chaussées, Paris, France, session 1841–1842, 223 pages (in French).

Bombardelli, F.A. (2012). Computational Multi-Phase Fluid Dynamics to Address Flows past Hydraulic Structures. *Proc. 4th IAHR International Symposium on Hydraulic Structures,* APRH – Associação Portuguesa dos Recursos Hídricos (Portuguese Water Resources Association), J. Matos, S. Pagliara & I. Meireles Eds., 9–11 February 2012, Porto, Portugal, Paper 2, 19 pages (CD-ROM).

Bradley, J.N., and Peterka, A.J. (1957a). The Hydraulic Design of Stilling Basins: Hydraulic Jumps on a Horizontal Apron (Basin I). *Journal of Hydraulic Division,* ASCE, Vol. 83, No. HY5, pp. 1401-1/1401-22.

Bradley, J.N., and Peterka, A.J. (1957b). The Hydraulic Design of Stilling Basins: High dams, Earth Dams and Large Canal Structures (Basin II). *Journal of Hydraulic Division,* ASCE, Vol. 83, No. HY5, pp. 1402-1/1402-14.

Bradley, J.N., and Peterka, A.J. (1957c). The Hydraulic Design of Stilling Basins: Short Stilling Basin for Canal Structures, Small Outlet Works and Small Spillways (Basin III). *Journal of Hydraulic Division,* ASCE, Vol. 83, No. HY5, pp. 1403-1/1403-22.

Bradley, J.N., and Peterka, A.J. (1957d). The Hydraulic Design of Stilling Basins: Stilling basin and wave Suppressors for Canal Structures, Outlet Works and Diversion Dams (Basin IV). *Journal of Hydraulic Division*, ASCE, Vol. 83, No. HY5, pp. 1404-1/1404-20.

Bradley, J.N., and Peterka, A.J. (1957e). The Hydraulic Design of Stilling Basins: Stilling Basin with Sloping Apron (Basin V). *Journal of Hydraulic Division*, ASCE, Vol. 83, No. HY5, pp. 1405-1/1405-32.

Bradley, J.N., and Peterka, A.J. (1957f). The Hydraulic Design of Stilling Basins: Small Basins for Pipe or Open Channel Outlets – No Tail water Required (Basin VI). *Journal of Hydraulic Division*, ASCE, Vol. 83, No. HY5, pp. 1406-1/1406-17.

Bung, D.B., Sun, Q., Meireles, I., Matos, J., and Viseu, T. (2012). USBR type III stilling basin performance for steep stepped spillways, *Proceedings of the 4th IAHR International Symposium on Hydraulic Structures*, APRH – Associação Portuguesa dos Recursos Hídricos, J. Matos, S. Pagliara & I. Meireles Eds., 9–11 February 2012, Porto, Portugal, Paper 4, 8 pages (CD-ROM).

BUREC (1980). *Hydraulic Design of Stilling Basins and Energy Dissipators*. Water Resources Publications, U.S. Dept. of Interior, Bureau of Reclamation, Denver.

Campbell. F.B., and Pickett, E.B. (1959). The Value of Prototype tests to Hydraulic Design. *Proc. 7th Hydraulics Conference*, 16–19 June 1958, Iowa, USA, edited by A. Toch and G.R. Scheider, State University of Iowa Publ., pp. 273–300.

Carvalho R.F. (2002). *Hydrodynamic actions in hydraulic structures: Numerical model of the hydraulic jump*. Ph.D. thesis, University of Coimbra, Portugal (in Portuguese).

Carvalho R.F., Lemos C.M., and Ramos C.M. (2008). Numerical computation of the flow in hydraulic jump stilling basins. *Journal of Hydraulic Research*, IAHR, Vol. 46, No. 6, pp. 739–775 (DOI: 10.3826/jhr.2008.2726).

Carvalho, R.F., and Martins, R. (2009). Stepped spillway with hydraulic jumps: scale and numerical models of a conceptual prototype. *Journal of Hydraulic Engineering*, ASCE, Vol. 135, No. 7, pp. 615–619 (DOI: 10.1061/_ASCE_HY.1943-7900.0000042).

Carvalho, R.F., and Lorena, M. (2012). Roughened Channels with Cross Beams Flow Features. *Journal of Irrigation and Drainage Engineering*, ASCE, Vol. 138, No. 8, pp. 748–756 (DOI: 10.1061/(ASCE)IR.1943-4774.0000457).

Chachereau, Y., and Chanson, H. (2011a). Free-Surface Fluctuations and Turbulence in Hydraulic Jumps. *Experimental Thermal and Fluid Science*, Vol. 35, No. 6, pp. 896–909 (DOI: 10.1016/j.expthermflusci.2011.01.009).

Chachereau, Y., and Chanson, H. (2011b). Bubbly Flow Measurements in Hydraulic Jumps with Small Inflow Froude Numbers. *International Journal of Multiphase Flow*, Vol. 37, No. 6, pp. 555–564 (DOI: 10.1016/j.ijmultiphaseflow.2011.03.012).

Chanson, H. (1997). *Air Bubble Entrainment in Free-Surface Turbulent Shear Flows*. Academic Press, London, UK, 401 pages.

Chanson, H. (2004). *The Hydraulics of Open Channel Flow: An Introduction*. Butterworth-Heinemann, 2nd edition, Oxford, UK, 630 pages.

Chanson, H. (2007). Bubbly Flow Structure in Hydraulic Jump. *European Journal of Mechanics B/Fluids*, Vol. 26, No. 3, pp. 367–384 (DOI:10.1016/j.euromechflu.2006.08.001).

Chanson, H. (2009a). Current Knowledge In Hydraulic Jumps And Related Phenomena. A Survey of Experimental Results. *European Journal of Mechanics B/Fluids*, Vol. 28, No. 2, pp. 191–210 (DOI: 10.1016/j.euromechflu.2008.06.004).

Chanson, H. (2009b). Development of the Bélanger Equation and Backwater Equation by Jean-Baptiste Bélanger (1828). *Journal of Hydraulic Engineering*, ASCE, Vol. 135, No. 3, pp. 159–163 (DOI: 10.1061/(ASCE)0733-9429(2009)135:3(159)).

Chanson, H. (2010). Convective Transport of Air Bubbles in Strong Hydraulic Jumps. *International Journal of Multiphase Flow*, Vol. 36, No. 10, pp. 798–814 (DOI: 10.1016/j.ijmultiphaseflow.2010.05.006).

Chanson, H. (2012). Momentum Considerations in Hydraulic Jumps and Bores. *Journal of Irrigation and Drainage Engineering*, ASCE, Vol. 138, No. 4, pp. 382–385 (DOI 10.1061/(ASCE)IR.1943-4774.0000409).

Chanson, H., and Brattberg, T. (2000). Experimental Study of the Air-Water Shear Flow in a Hydraulic Jump. *International Journal of Multiphase Flow*, Vol. 26, No. 4, pp. 583–607.

Chanson, H., and Montes, J.S. (1995). Characteristics of Undular Hydraulic Jumps. Experimental Apparatus and Flow Patterns. *Journal of Hydraulic Engineering*, ASCE, Vol. 121, No. 2, pp. 129–144. Discussion: Vol. 123, No. 2, pp. 161–164 (ISSN 0733-9429).

Crank, J. (1956). *The Mathematics of Diffusion*. Oxford University Press, London, UK.

Deng, Z., Guensch, G.R., Richmond, M.C., Weiland, M.A., and Carlson, T.J. (2007). Prototype measurements of pressure fluctuations in The Dalles Dam stilling basin. *Journal of Hydraulic Research*, IAHR, Vol. 45, No. 5 pp. 674–678.

Drew, D., and Prassman, S. (1999). *Theory of Multicomponent Fluids*. Springer, Applied Mathematical Sciences, Vol. 135.

Eriksson, J.G., Karlsson, R.I., and Persson, J. (1998). An experimental study of a two-dimensional plane turbulent wall jet. *Experiments in Fluids*, Vol. 25, pp. 50–60.

Ervine, D.A. (1998). Air Entrainment in Hydraulic Structures: a Review. *Proc. Instn Civ. Engrs, Water, Maritime & Energy*, UK, Vol. 130, pp. 142–153.

Ferziger, J.H., and Perić, M. (1996). Computational Methods for Fluid Dynamics. Springer-Verlag, Berlin, Germany, 356 pages.

FHWA (2006). *Hydraulic Engineering Circular 14 – "Energy Dissipators"*. U.S. Department of Transportation, Publication No. FHWA-NHI-06-086, Federal High-Way Administration.

Fiorotto, V., and Rinaldo, A. (1992). Fluctuating uplift and lining design in spillway stilling basins. Journal of Hydraulic Research, Vol. 38, No. 5, pp. 64–82.

Graham J.R., Creegan, P.J., Hamilton, W.S., Hendrickson, J.G., Kaden, R.A., McDonald, J.E., Noble, G.E., and Schrader, E. (1998). *Erosion of Concrete in Hydraulic Structures*. American Concrete Institute, Report by ACI Committee 210, ACI210R-93, 24 pages.

George, W.K., Abrahamsson, H., Eriksson, J., Karlsson, R.I., Lofdahl, L., and Wosnik, M. (2000). A Similarity Theory for the Turbulent Plane Wall Jet without External Stream. *Journal of Fluid Mechanics*, Vol. 425, pp. 367–411.

Habib, E., Mossa, M., and Petrillo, A. (1994). Scour Downstream of Hydraulic Jump. *Proc. Conf. Modelling, Testing & Monitoring for Hydro Powerplants*, Intl Jl Hydropower & Dams, Budapest, Hungary, pp. 591–602.

Hager, W.H. (1992). *Energy Dissipators and Hydraulic Jump*. Kluwer Academic Publ., Water Science and Technology Library, Vol. 8, Dordrecht, The Netherlands, 288 pages.

Henderson, F.M. (1966). *Open Channel Flow*. MacMillan Company, New York, USA.

Hirt, C.W. (2003). Modeling Turbulent Entrainment of Air at a Free Surface. *Flow Science*, FSI-03-TN61.

Hirt, C.W., and Nichols, B.D. (1981). VOF Method for the Dynamics of Free Boundaries. *Journal of Computational Physics*, 39, 201–225.

Hirt, C.W., and Sicilian, J. (1985). A porosity technique for the definition of obstacles in rectangular cell meshes. *Proceedings of 4th International Conference on Ship Hydrodynamics*, Washington, DC, September.

Hoyt, J.W., and Sellin, R.H.J. (1989). Hydraulic Jump as 'Mixing Layer'. *Jl of Hyd. Engrg.*, ASCE, Vol. 115, No. 12, pp. 1607–1614.

Ivoilov, A.A. (1982). Pressure fluctuations on the bottom of open channel turbulent flows in spillways structures. *Hydrotechnical Construction*, No. 10, Outubro.

Katz, Y., Horev, E., and Wygnanski, I. (1992). The Forced Turbulent Wall Jet. *Journal of Fluid Mechanics*, Vol. 242, pp. 577–609.

Kucukali, S., and Chanson, H. (2008). Turbulence Measurements in Hydraulic Jumps with Partially-Developed Inflow Conditions. *Experimental Thermal and Fluid Science*, Vol. 33, No. 1, pp. 41–53 (DOI: 10.1016/j.expthermflusci.2008.06.012).

Leandro, J., Carvalho, R., Chachereau, Y., and Chanson, H. (2012). Estimating Void Fraction in a Hydraulic Jump by Measurements of Pixel Intensity. *Experiments in Fluids*, Vol. 52, No. 5, Page 1307–1318 (DOI: 10.1007/s00348-011-1257-1).

Lemos, F.O. (1983). *Hydraulic structures research needs*. LNEC, Lisboa, Portugal (in Portuguese).

Lemos, C.M. (1992). *Wave Breaking – A Numerical Study*. Lecture Notes in Engineering, Edited by Brebbia and S.A. Orszag, Springer-Verlag, Vol. 71.

Leutheusser, H.J., and Schiller, E.J. (1975). Hydraulic Jump in a Rough Channel. *Water Power & Dam Construction*, Vol. 27, No. 5, pp. 186–191.

Liggett, J.A. (1994). *Fluid Mechanics*. McGraw-Hill, New York, USA.

Lighthill, J. (1978). *Waves in Fluids*. Cambridge University Press, Cambridge, UK, 504 pages.

Liovic, P., and Lakehal, D., (2012). Subgrid-scale modelling of surface tension within interface tracking-based Large Eddy and Interface Simulation of 3D interfacial flows. *Computers & Fluids*, Vol. 63, pp. 27–46.

Long, D., Rajaratnam, N., Steffler, P.M., and Smy, P.R. (1991). Structure of Flow in Hydraulic Jumps. *Jl of Hyd. Research*, IAHR, Vol. 29, No. 2, pp. 207–218.

Lubin, P., Glockner, S., Kimmoun, O., and Branger, H. (2011). Numerical study of the hydrodynamics of regular waves breaking over a sloping beach. *European Journal of Mechanics – B/Fluids*, Vol. 30, No. 6, pp. 552–564.

Ma, J., Oberai, A., Drew, D., Lahey Jr, R., Drew, Donald. (2011). Modeling Air Entrainment and Transport in a Hydraulic Jump using Two-Fluid RANS and DES Turbulence Models. *Heat Mass Transfer*, Vol. 47, No. 8, pp. 911–919.

Mccorquodale J.A., and Khalifa, A. (1983). Internal flow in hydraulic jumps. *Journal of Hydraulic Engineering*, ASCE, Vol. 109, pp. 684–701.

Morris, H. (1968). Hydraulics of energy dissipation in steep rough channels. *Research Division Bulletin*, Virginia Polytechnic Institute, Blacksburg VA, USA.

Mossa M., and Tolve, U. (1998). Flow visualization in bubbly two-phase hydraulic jump. *Journal of Fluids Engineering*, ASME, Vol. 120, pp. 160–165.

Mouaze, D., Murzyn, F., and Chaplin, J.R. (2005). Free Surface Length Scale Estimation in Hydraulic Jumps. *Journal of Fluids Engineering*, Trans. ASME, Vol. 127, pp. 1191–1193.

Murzyn, F., Mouaze, D., and Chaplin, J.R. (2005). Optical Fibre Probe Measurements of Bubbly Flow in Hydraulic Jumps *International Journal of Multiphase Flow*, Vol. 31, No. 1, pp. 141–154.

Murzyn, F., and Chanson, H. (2009a). Free-Surface Fluctuations in Hydraulic Jumps: Experimental Observations. *Experimental Thermal and Fluid Science*, Vol. 33, No. 7, pp. 1055–1064 (DOI: 10.1016/j.expthermflusci.2009.06.003).

Murzyn, F., and Chanson, H. (2009b). Experimental Investigation of Bubbly Flow and Turbulence in Hydraulic Jumps. *Environmental Fluid Mechanics*, Vol. 9, No. 2, pp. 143–159 (DOI: 10.1007/s10652-008-9077-4).

Novak, P., and Cabelka, J. (1981). *Models in Hydraulic Engineering. Physical Principles and Design Applications*. Pitman Publ., London, UK.

Pagliara, S., Lotti, I., and Palermo, M. (2008). Hydraulic Jump on Rough Bed of Stream Rehabilitation Structures. *Jl of Hydro-Environment Research*, Vol. 2, No. 1, pp. 29–38.

Peterka, A.J. (1978). *Hydraulic Design of Stilling Basins and Energy Dissipators*. United States Department of the Interior, Bureau of Reclamation, Engineering Monograph No. 25, Denver, Colorado, USA, 4th printing and revision, 240 pages.

Pinheiro, A.N., Quintela A.C., and Ramos C.M. (1994). Hydrodynamic Forces in Hydraulic Jump Stilling Basins. *Proc. Fundamentals and advancements in hydraulic measurements and experimentation Symposium*, Fundamentals and advancements in hydraulic measurements and experimentation, ASCE, C.A. Pugh Editor, pp. 321–330.

Pope, S.B. (2000). *Turbulent Flows*. Cambridge University Press, 771 pages.

Preobrazhenskii, N.A. (1958). Laboratory and field investigations of flow pressure pulsation and vibration of large dams. *Proc. 6th ICOLD Congress*, Question No. 21, R. 110, pp. 1 23.

Qingchao, L., and Drewes, U. (1994). Turbulence characteristics in free and forced hydraulic jumps. Journal of Hydraulic Research, Vol. 32, No. 6, pp. 877–898.

Rajaratnam, N. (1965). The Hydraulic Jump as a Wall Jet. *Jl of Hyd. Div.*, ASCE, Vol. 91, No. HY5, pp. 107–132. Discussion: Vol. 92, No. HY3, pp. 110–123 & Vol. 93, No. HY1, pp. 74–76.

Rajaratnam, N. (1967). Hydraulic Jumps. *Advances in Hydroscience*, Ed. V.T. CHOW, Academic Press, New York, USA, Vol. 4, pp. 197–280.

Rajaratnam, N. (1976). *Turbulent Jets*. Elsevier Scientific, Development in Water Science, 5, New York, USA.

Rao, N.S.L., and Kobus, H.E. (1971). *Characteristics of Self-Aerated Free-Surface Flows*. Water and Waste Water/Current Research and Practice, Vol. 10, Eric Schmidt Verlag, Berlin, Germany.

Resch, F.J., and Leutheusser, H.J. (1972). Le Ressaut Hydraulique: Mesure de Turbulence dans la Région Diphasique. ('The Hydraulic Jump: Turbulence Measurements in the Two-Phase Flow Region.') *Jl La Houille Blanche*, No. 4, pp. 279–293 (in French).

Richard, G.L., and Gavrilyuk, S.L. (2013). The Classical Hydraulic Jump in a Model of Shear Shallow-Water Flows. *Journal of Fluid Mechanics*, Vol. 725, pp. 492–521.

Rostamy, N., Bergstrom, D.J., and Sumner, D. (2009). An Experimental Study of a Turbulent Wall Jet on Smooth and Rough Surface. *Proc.Symp. on the Physics of Wall-Bounded Turbulent Flows on Rough Walls*, IUTAM, 6 pages. (Also: IUTAM Book Series 22, Springer, 2010, pp. 55–60 (DOI: 10.1007/978-90-481-9631-9_8)).

Rouse, H., Siao, T.T., and Nagaratnam, S. (1959). Turbulence Characteristics of the Hydraulic Jump. *Transactions*, ASCE, Vol. 124, pp. 926–950.

Ryu Y., Chang K.-A, and Lim H.-J. (2005). Use of bubble image velocimetry for measurement of plunging wave impinging on structure and associated greenwater. *Measurement Science and Technology*, Vol. 16, pp. 1945–1953.

Schwarz, W.H., and Cosart, W.P. (1961). The Two-Dimensional Wall-Jet. *Journal of Fluid Mechanics*, Vol. 10, Part 4, pp. 481–495.

Tachie, M.F., Balachandar, R., and Bergstrom, D.J. (2004). Roughness Effects on Turbulent Plane Wall Jets in an Open Channel. *Experiments in Fluids*, Vol. 37, No. 2, pp. 281–292.

Teixeira, E.D. (2008). Scale effects on the predictions of extreme values on the bottom of hydraulic jump stilling basins. *Ph.D. thesis*, Universidade Federal do Rio Grande do Sul, Instituto de pesquisas hidráulicas, Brasil (in Portuguese).

Toso, J.W., and Bowers, E.C. (1988). Extreme pressures in hydraulic jump stilling basins. *Journal of Hydraulic Engineering*, ASCE, Vol. 114, No. 8, pp. 829–843.

Valiani, A. (1997). Linear and Angular Momentum Conservation in Hydraulic Jump. *Journal of Hydraulic Research*, IAHR, Vol. 35, No. 3, pp. 323–354.

Wood, I.R. (1991). *Air Entrainment in Free-Surface Flows*. IAHR Hydraulic Structures Design Manual No. 4, Hydraulic Design Considerations, Balkema Publ., Rotterdam, The Netherlands, 149 pages.

Wygnanski, I., Katz, Y., and Horev, E. (1992). On the Applicability of Various Scaling Laws to the Turbulent Wall Jet. *Journal of Fluid Mechanics*, No. 234, pp. 669–690.

Yuditskii, G. (1963). Hydrodynamic actions caused by dicharge in stilling basin. *Izvestiya VNIIG*, Vol. 65,72, 73,77 (In Russian, translated into Portuguese by LNEC-72 n° 442,535, 541,520).

Zhang, G., Wang, H., and Chanson, H. (2013). Turbulence and Aeration in Hydraulic Jumps: Free-Surface Fluctuation and Integral Turbulent Scale Measurements. *Environmental Fluid Mechanics*, Vol. 13, No. 2, pp. 189–204 (DOI: 10.1007/s10652-012-9254-3).

BIBLIOGRAPHY

Chanson, H., and Chachereau, Y. (2013). Scale Effects Affecting Two-Phase Flow Properties in Hydraulic Jump with Small Inflow Froude Number. *Experimental Thermal and Fluid Science*, Vol. 45, pp. 234–242 (DOI: 10.1016/j.expthermflusci.2012.11.014).

Chanson, H., and Gualtieri, C. (2008). Similitude and Scale Effects of Air Entrainment in Hydraulic Jumps." *Journal of Hydraulic Research*, IAHR, Vol. 46, No. 1, pp. 35–44.

Gonzalez. A.E., and Bombardelli, F.A. (2005). Two-Phase Flow Theoretical and Numerical Models for Hydraulic Jumps, including Air Entraimnent. *Proc. 31st IAHR Biennial Congress*, Seoul, Korea, B.H. Jun, S.I. Lee, I.W. Seo and G.W. Choi Editors (CD-ROM).

Hager, W.H., Bremen, R., and Kawagoshi, N. (1990). Classical Hydraulic Jump: Length of Roller. *Journal of Hydraulic Research*, IAHR, Vol. 28, No. 5, pp. 591–608.

Hager, W.H., and Li, D. (1992). Sill-controlled energy dissipator. *Journal of Hydraulic Research*, IAHR, Vol. 30, No. 2, pp. 165–181.

Magalhães, A.P. (1974). *Divergent structures for energy dissipation. Hydraulic jumps in rectangular and divergent stilling basins with horizontal bed.* Report, LNEC, Lisboa, Portugal (in Portuguese).

Montes, J.S., and Chanson, H. (1998). Characteristics of Undular Hydraulic Jumps. Results and Calculations. *Journal of Hydraulic Engineering*, ASCE, Vol. 124, No. 2, pp. 192–205.

Murzyn, F., and Chanson, H. (2008). Experimental Assessment of Scale Effects Affecting Two-Phase Flow Properties in Hydraulic Jumps. *Experiments in Fluids*, Vol. 45, No. 3, pp. 513–521 (DOI: 10.1007/s00348-008-0494-4).

Murzyn, F., Mouaze, D., and Chaplin, J.R. (2007). Air-Water Interface Dynamic and Free Surface Features in Hydraulic Jumps. *Journal of Hydraulic Research*, IAHR, Vol. 45, No. 5, pp. 679–685.

Rajaratnam, N. (1962). An Experimental Study of Air Entrainment Characteristics of the Hydraulic Jump. *Journal of Institution of Engineers India*, Vol. 42, No. 7, March, pp. 247–273.

Rajaratnam, N. (1968). Hydraulic Jumps on Rough Beds. *Trans. Engrg. Institute of Canada*, Vol. 11, No. A-2, May, pp. I-VIII.

Stokes, G.G. (1856). On the effect of Internal Friction of Fluids on the Motion of Pendulums. *Trans. Camb. Phil. Soc.*, Vol. 9, Part II, pp. 8–106.

Ubbink, O., (1997). Numerical prediction of two fluid systems with sharp interfaces. *Ph.D. thesis*, Dept. of Mechanical Engineering, Imperial College of Science, Technology and Medicine, London, UK.

Chapter 5

Ski jumps, jets and plunge pools

Michael Pfister & Anton J. Schleiss
Laboratory of Hydraulic Constructions (LCH), Ecole Polytechnique
Fédérale de Lausanne (EPFL), Station 18, Lausanne, Switzerland

ABSTRACT

Ski jumps are frequent spillway elements conveying floods with high discharges and heads. The flow is separated from the chute and travels as a free jet, to finally impinge onto a plunge pool water cushion. During the flight phase, the jet disperses and disintegrates up to certain level, so that its energy input into the plunge pool is reduced. The dissipation of the residual energy occurs within the plunge pool. The design of a ski jump and the estimation of the plunge pool shape is challenging and of outstanding relevance, because the spillways is a safety element of a dam and the energy to dissipate can achieve several Gigawatts. The present chapter gives an overview of typical bucket types and their effect on the jet behavior, as well as its footprint shape when impinging on the plunge pool. Furthermore, aspects of plunge pool stability are discussed.

5.1 INTRODUCTION

5.1.1 Use of ski jumps

Ski jumps as outlet structures of chutes are of interest to convey floods with high energy heads and discharges respecting a safe distance from the dam, given that geology and topography allow their installation (Khatsuria 2005, Novak et al. 2007). They dissipate efficiently the residual jet impact energy in a plunge pool, far away from dam foundation and abutments. Ski jumps can also influence the jet direction and travel length. Consequently, the location and shape of the jet footprint impacting into the plunge pool and on the rock interface can be influenced, at least within certain limits. Besides, ski jumps remove the high-speed flow from the chute, sparing the latter further downstream from cavitation risk or strong pressure fluctuations.

A principal disadvantage of ski jumps and the corresponding jets is spray formation, leading to saturation and possible slides of adjacent valley flanks, and – in cold regions – freezing on roads and at switchyards. Further, residual impact energy can produce scouring of the mobile river bed and of the adjacent rock. Chutes combined with ski jumps should also be applied only for sediment free flows to avoid abrasion on the chute and flip bucket.

Ski jump spillways typically include an inlet structure, followed by a steep chute or a low level outlet gallery conveying the free surface flow. At the beginning of the chute or gallery, either intakes with gates are installed or un-regulated inlets exist, consisting of free overflow crests. To allow for a sufficient jet travel length for all operational conditions, regulated intakes are often preferred.

To design all features of the flip bucket and the plunge pool for the Probable Maximum Flood (PMF) is normally not economical (Mason 1993, Schleiss 2002). More frequent floods, as, for instance, the 500-years flood for the ski jump geometry and the 100- to 200-years flood for the pre-excavation of the plunge pool are recommended. However, catastrophic damages endangering the dam should not occur even for PMF. To take advantage of a ski jump, the minimum jet take-off velocity should exceed some 20 m/s (Heller et al. 2005). For unit discharges above 150 m²/s the cavitation risk on the chute and also at the bucket may be significant (Peterka 1964). Nevertheless, in combination with chute aerators (Pfister and Hager 2010a, b), unit discharges above 200 m²/s are not rare today (Mason 1993, Schleiss 2002).

The take-off lip elevation is typically located at some 30 to 50% of the dam height to produce an optimum jet impact location, measured from the plunge pool tailwater as reference (Vischer and Hager 1998). Higher take-off lips (above some 50% of the dam height) may have a too small take-off velocity resulting in a reduced horizontal jet length, whereas lower lips (below some 30% of the dam height) result in short jets due to the small offset between the bucket lip and the plunge pool water surface. To ensure a free jet issuance, the take-off lip should always be located above the maximal plunge pool water level so that a submergence of the latter is avoided.

Geometrical skip jump deflection angles α (Fig. 5.1) between $\alpha = 15$ to $35°$ are given by Peterka (1964), maximally $30°$ are recommended by USBR (1974), Rao (1982) mentions values from 30 to $40°$, a typical spectrum of $\alpha = 20$ to $40°$ is proposed by Vischer and Hager (1998), and Mason (1993) gives a most widely adopted range of $\alpha = 30$ to $35°$. The angle ε between the lower jet surface and the concrete structure downstream of the lip should exceed some $40°$ to avoid local negative pressures on the concrete surface (e.g. $90°$ in Fig. 5.1, Mason 1993). The lip elevation and the deflection angle are optimized to principally achieve an optimal jet travel length with an acceptable impact location.

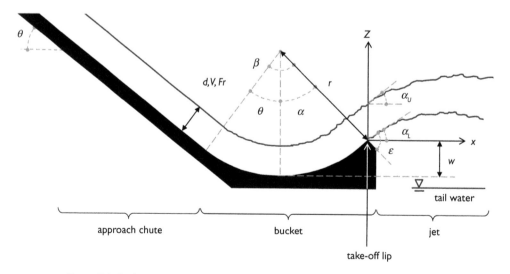

Figure 5.1 Definition sketch with geometry and flow parameters of a ski jump.

The curvature radius r of the bucket is affected by three aspects: (1) construction cost favoring small values of r resulting in small bucket foundation dimensions, (2) dynamic invert pressures forces opting for large r, and (3) the flow characteristics at the take-off lip, influencing flow choking, the effective jet take-off angle and turbulent features along the jet. Peterka (1964) and Mason (1993) suggest bucket radii of $r/d \geq 5$ for the design discharge, to avoid internal shearing or "tumbling", where d is the local flow depth upstream of the bucket (Fig. 5.1). Listing the characteristics of various prototype ski jumps indicates typical values of 12 m $< r <$ 19 m, with an average around $r = 15$ m (Mason 1993, Rao 1982). Rao (1982) suggests that $r/d = 4Fr - 15$ is optimal for $5.5 < Fr < 10$. Several other proposals to derive the bucket radius are summarized by Vischer and Hager (1995), also including these of USCE (1977) and USBR (1974), limiting the maximal flow-induced pressures on the bucket invert.

The above mentioned Froude number Fr is defined at the section immediately upstream of the bucket (Fig. 5.1) as

$$Fr = \frac{V}{\sqrt{gd}} \tag{5.1}$$

There, V = upstream flow velocity, and g = acceleration of gravity.

In the following, this chapter refers to free ski jumps without any submergence of the bucket resulting from the tailwater.

5.1.2 Ski jump types

An appropriate bucket geometry has to be defined depending on the desired jet trajectory and its footprint features. Some of the most frequent ski jump types are shown in Figure 5.2. There, "bucket type" refers to the longitudinal and transversal arrangement of the bucket, whereas "chute end layout" (Fig. 5.3) describes the geometry of the sidewalls in plan view. The sidewalls may not only vary exclusively at the bucket, but already upstream to avoid hydraulic phenomena such as flow separations or shockwaves at expansions or contractions. Generally speaking,

- the height difference between the reservoir water level and the take-off lip elevation has an influence on the initial take-off velocity, and thus on the jet travel length, at least as long as uniform flow conditions are not achieved on the chute,
- the height difference between the take-off lip elevation and the plunge pool water level has an influence on the jet travel length,
- the bucket radius affects mainly the effective take-off angle and the jet turbulence (linked to the disintegration process), influencing the jet travel length and its specific impact energy,
- the transversal inclination of the lip may result in a transversally-variable jet travel length and partially in a slight deviation in plan view. Then, an asymmetrical footprint occurs better fitting into a narrow valley,
- local inserts and baffles in the bucket or at the take-off lip affect mainly the initial jet turbulence and support its disintegration process, reducing thereby the energy impact into the plunge pool, and

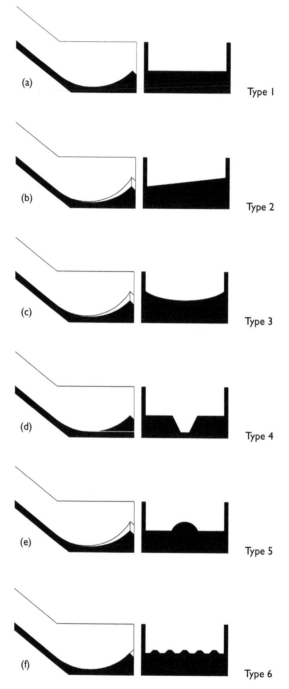

Figure 5.2 Sketches of bucket Types (a) 1, (b) 2, (c) 3, (d) 4, (e) 5, and (f) 6, shown as longitudinal section along the final chute part as well as transversally along the bucket take-off lip (view in downstream direction).

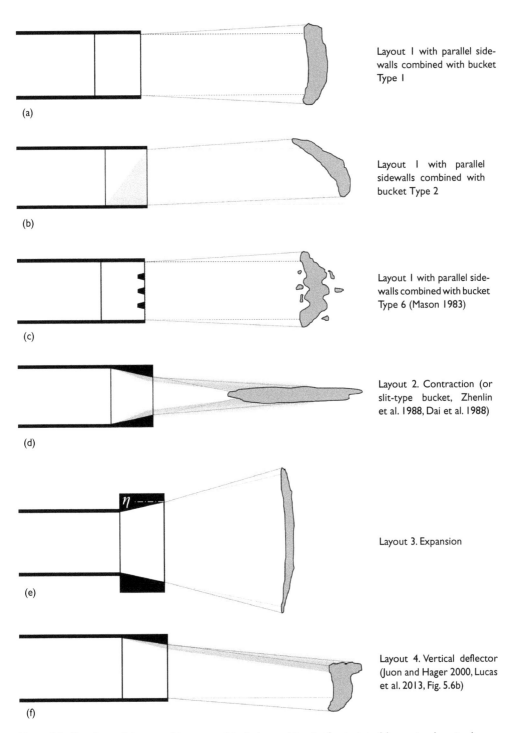

(a) Layout 1 with parallel side-walls combined with bucket Type 1

(b) Layout 1 with parallel sidewalls combined with bucket Type 2

(c) Layout 1 with parallel side-walls combined with bucket Type 6 (Mason 1983)

(d) Layout 2. Contraction (or slit-type bucket, Zhenlin et al. 1988, Dai et al. 1988)

(e) Layout 3. Expansion

(f) Layout 4. Vertical deflector (Juon and Hager 2000, Lucas et al. 2013, Fig. 5.6b)

Figure 5.3 Sketches of chute end Layouts with their resulting jet footprints (shown in plan view).

- the chute end width, possibly provided by an expansion or a contraction, determines the initial jet thickness which is linked to the disintegration process and the jet travel length.

Optimizing a flip bucket exclusively regarding the jet energy dissipation may result in high dynamic forces on the bucket, flow choking in the latter up to relatively large discharges, and the occurrence of pronounced shockwaves. An optimal combination of the bucket type and the chute end layout allows definition of an appropriate impact location of the jet into the plunge pool with a relatively high degree of disintegration. Some typical bucket types are (Fig. 5.2):

- Bucket Type 1 with transversally horizontal take-off lip, representing the standard configuration which results in a straight and relatively compact jet (Fig. 5.4a).
- Bucket Type 2 with transversally inclined take-off lip generating a relatively long jet trajectory at the steep side (small bucket radius) and a relatively short one at the flat side (large bucket radius), besides a slight lateral deviation (Fig. 5.5).
- Bucket Type 3 includes a transversally-rounded take-off lip, so that the jet take-off angle is steep near the sidewalls (for a concave lip curvature) but flat in the bucket center. Consequently, relatively long lateral trajectories are produced near the sidewalls, and shorter ones in the center. The vertical jet spread is reduced and the footprint more concentrated. A convex curvature reduces the lateral take-off angles and thereby shortens these trajectories, resulting in a rounded footprint area.
- Bucket Type 4 with a drainage channel improves its performance under small discharges by avoiding choking. The hydraulic jump in the bucket due to the counter slope does not occur for relatively small discharges, so that a free jet emerges. This type is mainly an option for unregulated spillways.
- Bucket Type 5 with insert(s) enhances the jet disintegration and the footprint area. The insert(s) also produces additional turbulence in the jet and lifts the central jet portion (Fig. 5.6a).
- Bucket Type 6 with baffles roughens the lower jet trajectory and enhances the flow turbulence in the lower part of the jet, enhancing its disintegration process. The footprint area is enlarged and thus the density of the impact energy reduced. Furthermore, the footprint is, at least up to certain discharges, not coherent but consists of several individual patches (Fig. 5.7a).

In addition to the bucket types as presented in Figure 5.2, also the chute end layout with the sidewall configuration strongly affects the shape and the surface area of the jet and its footprint. The standard configuration with parallel sidewalls is shown in Figure 5.3a as Layout 1, including a bucket Type 1. The jet spreads transversally due to the sidewall-induced turbulence (Ervine and Falvey 1987) and shows a slightly rounded footprint at the impact. This shape is initiated by the flow velocity distribution at the take-off lip. Near the bottom and at the sidewalls, the local flow velocity is in the order of 80% of the average value (Chow 1959). Accordingly, the resulting lateral jet trajectories are shorter as compared to those in the center, where the local velocity corresponds to about 120% of the average value.

Figure 5.4 Spillway of (a) Venda Nova Dam (Portugal, Photo Michael Pfister), and (b) Wivenhoe Dam (Australia) during operation (Photo Hubert Chanson), both with buckets Types 1 according to Figure 5.2.

Figure 5.5 Principal spillway of Zipingpu Dam (China) with a bucket Type 2 (Photo Michael Pfister).

The footprint shape is influenced by varying the bucket take-off lip and its angle, whereas the flow velocity distribution is determined by the chute and cannot be changed. An example of this concept is shown in Figure 5.3b, combining the chute end Layout 1 with a bucket Type 2. A small geometrical take-off angle occurs at the left (large bucket radius with low take-off elevation), and a high take-off angle at the right. Accordingly, a shorter jet trajectory emerges at the left and a longer at the right, so that the footprint consequently is rotated. Additionally, the jet is slightly deviated to the left due to the lateral acceleration initiated by the bucket. The slightly rounded footprint remains due to the aforementioned velocity distribution initiated by the chute. It has to be noted that a similar footprint may be achieved with an oblique take-off lip in plan view (Berchtold and Pfister 2011, Fig. 5.7a). The combination of a bucket Type 6 with the chute end Layout 1 is shown in Figure 5.3c. The baffles add turbulence to the jet, enhancing its disintegration process. Accordingly, the footprint is spread over a larger area and partially discontinuous.

Basically, all bucket Types from Figure 5.2 can be combined with the chute end Layouts 1 to 4 shown in Figure 5.3. Particularly, contracting or expanding sidewalls are often used. They either reduce the jet width and increase thereby its vertical thickness (Fig. 5.3d), or widen the jet transversally expanding its footprint (Fig. 5.3e). The individual widening angle η per sidewall should not exceed the value $\tan\eta = 1/(3Fr)$ to avoid flow separation from the wall (USBR 1974). As a consequence, the widening often starts upstream of the bucket (Berchtold and Pfister 2011, Fig. 5.7). For contractions (Fig. 5.3d), shockwaves are generated at the beginning of the lateral deflectors

(a)

(b)

Figure 5.6 Alqueva Dam (Portugal) spillway with (a) an insert at the take-off lip (left spillway, bucket Type 5), and (b) a vertical deflector on the bucket of the right spillway (Layout 4, both Photos Michael Pfister).

Figure 5.7 Chute end of Kárahnjúkar Dam spillway (Iceland, plain end-overfall with oblique take-off lip equipped with baffles) with (a) roll-waves initiated by small a discharge, and (b) in operation with a discharge of some 300 m³/s (Photos Landsvirkjun, Iceland).

(Vischer and Hager 1998). Given that the deflectors represent a positive abrupt wall deflection, positive shockwaves with a relatively increased flow depth are created. These represent zones with a comparatively high unit discharge, continuing or spreading along the jet and producing a flow concentration at the footprint. The contraction should be carefully designed in order to avoid flow choking of the bucket.

For extreme cases, such as slit-type buckets, the left- and right-side shockwaves merge and the jet is straighten up (Fig. 5.3d), enhancing the vertical dispersion of the jet. They mainly increase the upper jet take-off angle resulting in a long and thin footprint area. For small discharges, such buckets concentrate the flow and create a relatively long jet, thereby reducing the spectrum of the jet travel length as a result of varying discharges (Fig. 5.8).

Of course, asymmetrical set-ups are also possible, as for instance shown in Figure 5.3f. The effect of the shockwave on the footprint has to be taken into account. Such a bucket diverts the jet laterally – at least up to a certain degree – away from the side where the deflector is installed. Another approach to affect the horizontal jet orientation is to transversally incline the invert in the ski jump (similar to Fig. 5.2b). Various other ski jump arrangements were built or at least model-tested, as, for instance, types with platforms or particular baffle shapes (Roose and Gilg 1973). It is even possible to combine a ski jump with a stilling basin, as reported by Peterka (1964). For particular situations, two similar jets colliding in the air may be created by two symmetrical ski jumps. Factorovic (1952) gives an equation of the energy which is dissipated by jet collision in the air as a function of the individual unit discharges and velocities. For example, about 50% of the kinetic energy is dissipated under a

Figure 5.8 Paradela Dam (Portugal) spillway including a narrow flip bucket (Layout 2) downstream of a long chute (Photo Michael Pfister).

horizontal impact angle of 90° for two identical jets. Other set-ups of jet collision are proposed by Lencastre (1985). From an operational point of view, the adequate handling of the gates to guarantee jet collision for all discharges is rather challenging.

5.2   SKI JUMP HYDRAULICS

5.2.1   Bucket with transversally horizontal take-off lip

5.2.1.1   Parameter definition

The standard ski jump (Figs. 5.2a and 5.3a) is composed of a rectangular approach chute of constant slope θ and width (Fig. 5.1), conveying a design discharge with a flow depth d, a flow velocity V, and a Froude number Fr (Eq. 5.1) defined closely upstream of the bucket entrance. The latter includes mostly a circular-shaped invert with a curvature radius r, resulting in a geometrical take-off angle α (defined relative to the horizontal). The total deviation angle is accordingly $\beta = \theta + \alpha$, and the bucket height relative to the horizontal becomes $w = r(1 - \cos\alpha)$.

The jet is defined within the horizontal/vertical coordinate system $(x;z)$ according to Figure 5.1, with the point of origin at the take-off lip. The effective jet take-off angles to consider for the trajectory computation are α_U and α_L, with subscript U standing for the upper trajectory, and L for the lower trajectory.

5.2.1.2   Bucket pressure

The deviation of a high-speed flow induces forces on the related structure. Ski jumps combine high-speed flow of large discharges with a pronounced flow deviation (large β and small r), so that the bucket is loaded by high dynamic pressures. As a consequence, buckets are massive concrete structures, as for instance shown in Figures 5.5 and 5.6. USCE (1977) indicates that the pressures increase with discharge and head (i.e. flow velocity), and reduces with augmenting r. Mason (1993) gives the sum of the hydrodynamic and hydrostatic relative bottom pressure head h_p/d on the bucket as

$$\frac{h_p}{d} = 1 + \frac{V^2}{gr} = 1 + \frac{q^2}{grd^2} \tag{5.2}$$

Juon and Hager (2000) derive the relative maximum (dynamic) bottom pressure head based on the bend number $Be = (d/r)^{1/2}Fr$, as

$$\frac{h_p}{d} = Be^2 \tag{5.3}$$

Heller et al. (2005) systematically measured pressures along the approach channel and the flip bucket in physical models using pressure taps. The model used only a horizontal chute, i.e. $\theta = 0°$. For this condition, they measured a local relative maximum dynamic pressure head of

$$\frac{h_p}{d} = \frac{1}{5}\frac{\alpha}{40°}Fr^2 \tag{5.4}$$

Accordingly, small deflection angles produce smaller maximum pressure heads than do large deflection angles. The location of the maximum pressure was found near the take-off lip. Kavianpour and Pourhasan (2005) report pressure fluctuations on buckets derived from the physical model of Salman Farsi Dam (Iran).

The increased invert pressure on the bucket is also transmitted to the toe of the sidewalls, which have to be designed to support this load. Rao (1982) estimates that these have to resist to about 85% of the bucket pressure at the invert, consisting of the hydrostatic plus the dynamic component.

5.2.1.3   Bucket flow choking at small discharges

For non-controlled spillway inlets, small discharges occur at the beginning and the end of a flood event (Roose and Glig 1973). If the take-off lip is located on a higher elevation than the lowest point of the bucket (negative slope) then the flow momentum may be insufficient to maintain supercritical flow all along the bucket. Consequently, a hydraulic jump occurs in the bucket (called *flow choking*) and the take-off lip operates as an end-sill similar to a stilling basin. The flow passing the sill falls close to the bucket foundations, similar to a overflown weir, endangering their stability by erosion. Moreover, this may cause saturation of the ground leading to landslides or frost in rock cracks. Thus, ski jumps are – if possible – designed to avoid choking at small discharges by providing a gated spillway intake. Otherwise, a small bucket radius r with a low geometrical take-off angle α may avoid choking, or a drainage channel according to bucket Type 4 (Fig. 5.2). Finally, stabilization measures with heavy concrete linings near the bucket foundations can be provided.

Flow choking on flip buckets can be described in principle by applying the momentum equations as for stilling basins. Heller et al. (2005) distinguished between the increasing and decreasing discharge regime when analyzing the phenomenon of choking. They identify a limit bucket height w (Fig. 5.1) above which choking is probable, for a certain discharge. These limit heights are (for $1 < Fr < 4$)

$$w = 0.6\,d\left(Fr - 1\right)^{1.2} \tag{5.5}$$

for the increasing discharge regime, as well as

$$w = 0.9\,d\left(Fr - 1\right)^{0.9} \tag{5.6}$$

for the decreasing regime. The equations allows the prediction of the upper and lower limit of Fr for the increasing and decreasing regimes, and thus also the limit discharges to avoid choking of the bucket by a hydraulic jump.

Another hydraulic phenomenon occurring at small discharges combined with long chutes of constant shape and slope is roll-waves (Hager 2002, Di Cristo et al. 2008).

They create non-steady discharge peaks on the bucket, if no choking occurs. Further, shockwaves, generated for instance at gate piers and at converging sidewalls, may generate transversal discharge concentrations, affecting the spatial jet characteristics and the unit energy impact into the plunge pool.

5.2.2   Buckets with particular take-off lip arrangements

The standard bucket configuration with parallel sidewalls (Figs. 5.2 and 5.3a) represents a relatively simple structure in terms of geometry and hydraulics. Regarding its hydraulic performance, it may not represent an optimal solution for a certain prototype, as it generates typically a relatively compact jet with a small footprint, introducing a high degree of residual energy on a concentrated surface into the plunge pool.

Adaptions of the take-off lip, the sidewalls and the addition of inserts or baffles enhance jet disintegration and consequently reduce the energy impact into the plunge pool. The footprint can be orientated at a desired location and a favorable shape is produced. Such particular buckets are mostly site-specific and challenging to define, so that physical model tests are normally required. Some concepts and their effect on the jet and the footprint are shown in Figures 5.2 and 5.3 and briefly explained hereafter.

Inserts and baffles are an efficient way to increase the initial turbulence of the jet and to favor its disintegration and spread, thereby finally reducing the maximum dynamic plunge pool bottom pressures. A disintegrated and spread jet occurs if the projection angle on the baffles is high (for example 40°). Then, the flow between the baffles follows the less inclined bucket invert, and the flow on the baffles follows their steeper surface (Fig. 5.7b). Accordingly, the jet footprint on the plunge pool surface is stretched. However, small discharges will be "cut" into several jets according to the number of openings between the baffles. Mason (1983) finally recommends to aerate the downstream face of the baffles to avoid cavitation formation.

As an example, the 31 m wide chute end of the Kárahnjúkar Dam spillway (Iceland, Fig. 5.7) includes seven 3.8 m-spaced baffles with a width to height ratio of 0.5×0.5 m. The design flow depth at the chute end is about 1.5 m. The axes of the baffles are aligned parallel to the chute center line. To avoid cavitation damage, the side faces of the baffles are inclined by 5V:3H and their edges rounded (Minor 1988). The downstream faces of the baffle are aligned with the chute take-off lip. Berchtold and Pfister (2011) analyzed the time-averaged dynamic pressure heads at the plunge pool bottom resulting from configurations with and without baffles. The presence of the seven relatively small baffles reduced the dynamic pressures by 50%. It has to be noted that the sidewalls expand along a chute length of 125 m from 17 m to 31 m to avoid flow separation.

5.3   JET CHARACTERISTICS

5.3.1   Take-off angle

Laboratory experience shows that the jet take-off angles α_U and α_L (subscript U for upper and L for lower trajectory, Fig. 5.1) differ from the geometrical bucket angle

α (e.g. Orlov 1974, Dhillon et al. 1981, Vischer and Hager 1998, Heller et al. 2005, Pfister et al. 2014). Values of α_U and α_L follow from physical model tests, by fitting the effectively measured trajectories instead of considering local jet angles. Thus, the trajectory in its far-field is correctly represented, where the jet impact onto the plunge pool is expected. Such take-off angles α_U and α_L, defined relative to the horizontal (Fig. 5.1), are (Pfister et al. 2014)

$$\tan\alpha_U = 0.84\,\Lambda - 0.04 \tag{5.7}$$

$$\tan\alpha_L = 0.80\,\Lambda - 0.07 \tag{5.8}$$

These are referred to as α_j in Eq. (5.10), allowing for accurate trajectory prediction. The effects of the bucket geometry and the approach flow features are taken into account by the parameter (Pfister et al. 2014)

$$\Lambda = \tan\alpha\left(1 - \frac{d}{r}\right)^{1/3} \tag{5.9}$$

within the limits $-0.32 \leq \Lambda \leq 0.84$.

This approach includes also negative geometrical bucket angles ($-18° \leq \alpha \leq 40°$). For small approach flow depths d and large bucket radii r, the jet take-off angles α_j (i.e. α_U and α_L) are thus only slightly below the geometrical angle α, because the flow follows the bucket curvature. In contrast, large d and small r values result in α_j which are significantly smaller than α. Then, the flow is not able to precisely follow the bucket curvature anymore.

For positive α_j and $Fr < 4$, the jet does not follow the parabola near the take-off. Such conditions are close to bucket choking and must be avoided (Heller et al. 2005).

5.3.2   Trajectories

Peterka (1964) stated that "the spreading and trajectory characteristic of a given bucket can be determined only by testing in a hydraulic model". Nevertheless, the trajectories may be predicted with a sufficient accuracy at least for a standard bucket (USCE 1977). The knowledge of the precise jet trajectories is essential, as they define the jet impact location and angle in the plunge pool.

Particularly for non-gated spillways, the effectively-occurring discharges at the ski jump are gradual, reaching from zero discharge up to the design value. Accordingly, a wide spectrum of flow velocities and depths is observed at the flip bucket. The jet therefore covers the entire spectrum of trajectory length: from zero (for a choked flip bucket) up the travel length for the maximum discharge. Note that the water level rise in the plunge pool also affects the effective jet travel length. Given the typical shape of reservoir-outflow hydrographs, the small jump lengths (periods with small discharges) may last for a longer time than the maximum jet travel lengths.

The effective prototype jet trajectories and boundaries are challenging to define, given that the jet surface disintegrates under the influence of the initial flow turbulence producing a spreading jet. Following current standards in high-speed two-phase flow research, the virtual flow surface is usually defined where the local air concentration is $C = 0.90$. References published after the year 2000 mostly use such a criterion to derive the jet trajectory from laboratory testing.

The upper and lower jet trajectories may be approximated with the ballistic parabola as (Fig. 5.1)

$$z = x \tan\alpha_j - \frac{gx^2}{2V_j^2 \cos^2\alpha_j} \tag{5.10}$$

The subscript j refers to the effective jet take-off parameters (Eqs. 5.7 and 5.8), either for the upper or the lower trajectory. The lower trajectory starts at $(x;z) = (0\text{ m}, 0\text{ m})$, whereas the take-off point of the upper trajectory is typically at $(0\text{ m};1.3d)$ (Pfister et al. 2014).

Equation (5.10) describes the curve of a mass point of constant density under the influence of gravity, depending on the take-off conditions and neglecting jet disintegration as well as aerodynamic interaction. Thus, the computed trajectory represents a simplified theoretical approach, while effective trajectories observed on hydraulic models or prototypes may differ considerably from this assumption. The reason for this difference is based on uncertainties concerning the effective take-off velocity V, the effective flow depth d, the jet take-off angle α_j, and the disintegration processes combined with an interaction with the surrounding air. Consequently, the jet length in prototype is often shorter than derived by Eq. (5.10). USBR (1974) proposes a factor of 1.1 to multiply with the negative term of Eq. (5.10), i.e. to shorten the jet, in order to take into account these uncertainties. Gunko et al. (1965) observed a reduction of the trajectory length at $Fr > 6$. Kawakami (1973), who compared prototype data with the theoretical jet trajectories, observed an effect if $V > 13$ m/s.

If the approach flow along the chute upstream of the ski jump is fully aerated (Straub and Anderson 1958, Hager 1991, Chanson 1994, 1996), then its take-off velocity is higher than that of black-water flow. Prototype observations show that longer jets result for these conditions as compared to black-water approach flow (Minor 1987). For fully-aerated uniform flow conditions and chutes steeper than 20°, the drag is reduced under the influence of the air (Wood 1991, Chanson 1994). This circumstance must be taken into account when deriving the jet take-off velocity. Furthermore, the upper jet surface becomes rougher and more aerated, so that it is, by trend, above the theoretical trajectory for black-water approach flow (Pfister and Hager 2012). The lower jet trajectories are slightly reduced to about 80% of the reference trajectory.

The jet laterally spreads in the streamwise direction. The effective dispersion angle is difficult to determine and depends on the take-off conditions, particularly on the initial jet turbulence intensity (Ervine et al. 1997). The literature mentions values (per trajectory) of 2 to 5° for bucket type 1 (Fig. 5.1, Peterka 1964, Pfister et al. 2014), and values up to 12° for particular bucket shapes (Peterka 1964). The vertical dispersion is included in the computation of the trajectories, because $\alpha_U > \alpha_L$ if using Eqs. (5.8) and (5.9).

The jet impact angle on the plunge pool surface is obtained from the derivative of Eq. (5.10).

5.3.3 Air features

The expression "jet disintegration" is used to describe the continuous roughening process of the jet surfaces due to turbulence, starting at the take-off lip for the lower trajectory and slightly upstream of it for the upper trajectory. Jet disintegration comes along with aeration and jet dispersion.

According to Rajaratnam (1976), jet disintegration is initiated by: (1) turbulence resulting from chute flow and from jet take-off, (2) aerodynamic interaction between the jet and the surrounding air, and (3) relaxation of the velocity and pressure profiles. An analysis of the jet turbulence in the streamwise direction conducted by Ervine et al. (1995) indicates that (1) the turbulence intensity decreases, and (2) that the size of a turbulence cell increases, tending to cover the entire section. The jet breaks-up after 50 to 100 nozzle diameters (Ervine and Falvey 1987), where the coherent inner core completely disappears. This indicates that disintegration requires a certain developing length, or *vice versa*, it mainly takes place in the first jet reach. It causes rough and dispersed jet surfaces with entrapped air transport and, after a certain jet length, a fully aerated jet core (Fig. 5.9, Toombes and Chanson 2007, Schmocker et al. 2008). Hence, the density of the air-water mixture jet reduces with flow distance, and the interaction with the surrounding air increases. The resulting larger jet-footprint on the plunge pool surface, as well as the decreased density, can reduce the scour potential on a loose river bed (Pagliara et al. 2006, Canepa and Hager 2003). Such a loose bed is however rapidly washed out during operation. As for the remaining bare rock, disintegrated jets may increase the rock scour potential due to air entrainment into the plunge pool. Within aerated plunge pools, the water hammer velocity is reduced in the rock fissures, thus generating resonance phenomena, breaking-up the rock and ejecting boulders into the plunge pool by dynamic uplift (Bollaert and Schleiss 2003a, b).

The disintegration process of a jet is enhanced by (Vischer and Hager 1995):

- a high initial turbulence at the take-off (high approach flow velocity, large deflection angle, installation of baffles and lateral deflectors), and
- a non-circular jet cross-section with a small flow depth (as generated for instance by a chute end widening) to counter the compactness of the jet.

Figure 5.9 Jet features on a physical model, showing transition from black-water approach flow to fully aerated jet (Photo VAW, ETH Zurich).

As shown by Pfister and Hager (2009), the jet air concentration characteristics are determined by the black-water (subscript b) core length L_b (Fig. 5.10). It has to be noted that in Figure 5.10 the coordinate system was rotated, so that x' is now parallel to the chute bottom, again with the origin at the take-off lip. The black-water core length is defined with the minimum (subscript m) air concentration C_m measured within a certain jet air profile. Along L_b values $C_m < 0.01$ occur, whereas further downstream $C_m > 0.01$.

Pfister et al. (2014) derived from extensive model tests for a bucket Type 1 and a Layout 1 a black-water core length of ski jump jets as

$$\frac{L_b}{d} = 76\,Fr^{-1}\left(1+\tan\delta\right)^{-4}\left(1+\sin\theta\right) \tag{5.11}$$

with $\tan\delta = (1-\cos\beta)/\sin\beta$ as equivalent deflection angle. Based on L_b and the inclined coordinate x', the proposed streamwise normalization corresponds to a multiple of L_b as

$$\chi = \frac{x'}{L_b} \tag{5.12}$$

Then, $0 < \chi < 1$ corresponds to the black-water core portion, and $\chi > 1$ to the fully-aerated jet portion with $C_m > 0.01$. The average (subscript a) air concentration C_a per profile (parallel to the z' axis) is defined according to Straub and Anderson (1958) as an integration between both trajectories (at $C = 0.9$ for upper and lower trajectory). Then, the following values of C_a along χ were measured (Pfister et al. 2014)

$$C_a = \tanh\left(0.4\chi^{0.6}\right) \tag{5.13}$$

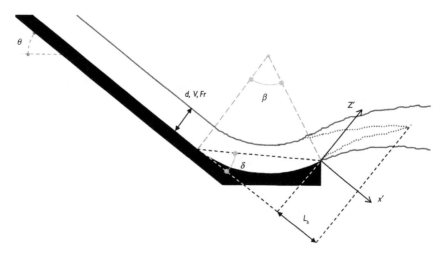

Figure 5.10 Definition sketch to derive the air-flow features of a ski jump generated jet.

The average jet air concentration at the jet impact location may be estimated with Eq. (5.13), in order to derive its potential to produce mean dynamic pressures acting on the plunge pool rock.

5.4 PLUNGE POOL

5.4.1 Water level drawdown

The impacting jet entrains flow momentum into the plunge pool. For relatively narrow valleys, the plunge pool water surface upstream of the impact location is lowered as compared to the level of the remaining (downstream) part. This drawdown is of relevance if the powerhouse is located near the dam toe regarding tailwater pressures at the turbines or concerning an inundation of the bottom outlet as well as regarding scour depths. Peterka (1964) reports a drawdown of almost 8 m at Hungry Horse Dam (USA). The drawdown at Karahnjukar Dam (Island) with a plunge pool in a narrow canyon reached, following model tests, maximally 5 m (Berchtold and Pfister 2011). Furthermore, the physical model of the Kariba Dam plunge pool revealed 3 to 3.5 m drawdown although the plunge pool is relatively wide (Bollaert et al. 2012).

5.4.2 Scour generating mechanisms

The scour extension at the plunge pool bottom is influenced by the characteristics of the falling jet, the plunge pool shape with its internal flow currents, and the surrounding rock mass. The physical processes are described by Schleiss (2002) as:

1. free falling jet behavior in the air and aerated jet impingement,
2. plunging jet behavior and turbulent flow in the plunge pool,
3. pressure fluctuation at the water-rock interface,
4. propagation of dynamic water pressures into rock joints,
5. hydrodynamic fracturing of closed ended rock joints and splitting of the rock,
6. ejection of rock blocks by dynamic uplift into the plunge pool (this includes removal of fractured rock block elements),
7. break-up of the rock blocks by the ball milling effect of the turbulent flow in the plunge pool, and
8. formation of a downstream mound and displacement of the scoured materials by sediment transport.

The entire process is – among others – time-dependent. The fractioning and removal of the rock requires a certain time period until an equilibrium scour shape is achieved. The plunge pool reaches its ultimate extension successively during this process, as the dynamic forces stressing the rock become smaller while the plunge pool volume increases, as the jet propagation length within the plunge pool augments during the process.

A physically-based scour model for the prediction of the ultimate scour depth of jointed rock was proposed by Bollaert et al. (2003a, b). It is based on three sub-modules: the falling jet, the plunge pool and the rock mass (Bollaert and Schleiss 2005, Schleiss and Annandale 2007). The rock mass module incorporates two sub-models expressing two failure criteria of the rock mass. The first one is the *Comprehensive Fracture Mechanics* (CFM) model, which determines the ultimate scour depth by expressing the instantaneous or time-dependent crack propagation due to hydrodynamic pressures. The second one is the *Dynamic Impulsion* (DI) model, describing the ejection of rock blocks from their mass. Each of these sub-models consists of a physical way to describe the destruction of the rock mass. The appropriate sub-model depends on the geomechanical characteristics of the rock, and more specifically on its degree of break-up.

5.4.3   Design flood for scour estimation

Spillways are mostly laid out for the design flood, typically a 1000-years flood, and their operation is checked for the safety flood, often PMF. Nevertheless, it is essential to estimate for which flood (and for which duration of the latter) the maximum scour depth occurs (during a certain period, for instance the technical lifetime of a dam), and on which flood event the constructive scour control measures have to be based. The ultimate scour depth will occur only if steady conditions are achieved in the scour hole, which is only the case after a long duration of spillway operation. It is thus conservative to base the scour depth estimation or the mitigation measures on low frequency floods (e.g. PMF or 1000-years flood). It appears unlikely that, during the technical lifetime of the dam, these rare floods can produce ultimate scour depths.

It seems reasonable to choose a design discharge with a probability of occurrence of about 50% during the lifetime of a dam for the scour evaluation and the protection measures, which is typically a 100 to 200-years flood. Higher design discharges with lower probability of occurrence are often too conservative. It has to be noted that, for gated free surface and orifice spillways, high discharges can be released at any time, also because of faulty operation. Furthermore, low-level outlet spillways generate a jet core impact velocity which is nearly independent from discharge.

5.4.4   Scour control measures

To limit damages as a consequence of scouring, three active measures are feasible (Schleiss 2002): To (1) avoid scour formation completely, (2) design the spillway so that the scour occurs far away from dam foundation and abutments, and (3) limit the scour extent. Since structures for scour control are rather expensive, normally only the two latter are economically viable (Ramos 1982). Besides elongating as much as possible the impact zone of the jet by an appropriate design of the ski jump, the extent of the scour can be influenced by the measures listed subsequently.

5.4.4.1   Limitation of unit discharge

This measure is efficient for arch dams and free ogee crest spillways with a jet impact location close to the dam toe. The unit discharge has to be reduced (i.e. the jet is widened) if the foundation is endangered by scour. Then – unfortunately – the take-off velocity at the lip and thereby the jet length are also reduced. For low-level outlets, the unit discharge depends on the size of the outlet openings. Since the velocity at gated spillways is high enough to divert the jet far away from the dam and its foundation, the limitation of the unit discharge is normally less important than for free crest spillways.

5.4.4.2   Enhancement of jet disintegration

Inserts, baffles, large take-off angles and high approach flow velocities at ski jumps enhance the jet disintegration process. All these measures also increase the air entrainment into the jet, principally reducing its scouring capacity. Martins (1973) suggested a reduction of 25% of the calculated scour depth for highly aerated jets, and 10% for intermediate air entrainment. Mason (1989) proposes an empirical expression considering the volumetric air-to-water ratio. On the other hand, the presence of air influences the water hammer velocity and thus the resonance effect of pressure waves in rock joints. Duarte (2014) observed that aerated jets reduce pressure fluctuations at the water-rock interface but may increase resonance effects in rock fissures. Although mostly favorable, forced aeration may thus increase under certain conditions the risk of rock block break-up and ejection.

5.4.4.3   Effect of tailwater dam

The depth of the plunge pool water cushion is increased by providing a tailwater dam. For plunge pool depths smaller than 4 to 6 times the jet diameter, core jet impact is normally observed on the plunge pool bottom (Bollaert 2002). Maximum fluctuations at the plunge pool bottom were observed for depths between 5 to 8 times the jet diameter (Bollaert and Schleiss 2003a, b). Substantial high values persist up to tailwater depths of 10 to 11 times the jet diameter. Therefore, only water cushions deeper than 11 times the jet diameter have a notable retarding effect on the scour formation.

5.4.4.4   Pre-excavation of plunge pool

Pre-excavation increases the tailwater depth, so that the same remarks can be made as for the tailwater dam. The pre-excavated plunge pool shape should approach the expected natural scour geometry for a loose rock. For the scour at Kariba Dam on Zambezi River, for instance, which reached a depth of more than 80 m below the tailwater level, the existing scour hole was reshaped by an excavation mainly in the downstream direction (Noret et al. 2013). This measure reduces significantly the dynamic pressures at the water-rock interface and guides the jet propagation across the plunge pool towards the downstream (Bollaert et al. 2012). If scour progresses deeper than the level of the pre-excavation at extreme floods, then the riverbank slopes have to be stabilized with pre-stressed rock anchors allowing for a certain undermining (Schleiss 2002).

5.4.4.5   Concrete lined plunge pools

If scour formation is unacceptable, then the rock surface in the plunge pool has to be reinforced and tightened by a concrete lining. Since the thickness of the lining is limited by construction and economic reasons, high tension or pre-stressed rock anchors are required to ensure the lining stability in view of the dynamic loading. Furthermore, the surface of the lining has to be protected with reinforced and high tensile concrete resistant against abrasion. Construction joints have to be equipped with water stops. Further, the stability of the lining against static uplift (during dewatering of the plunge pool) has to be assured by an adequate drainage system below the concrete slabs. The latter can also limit severe dynamic uplift for limited cracking of the lining (Mahzari et al. 2002).

As plunge pool linings represent a protection with a certain failure risk, a conservative design concept is necessary, considering:

- large-sized reinforced slabs, which average local instantaneous pressure fluctuations over a large area,
- water stops between slabs to prevent dynamic pressures penetrating underneath,
- anchor bars or pre-stressing cables, designed on the basis of the mean or instantaneous pressure differentials above and underneath the slabs, on the assumption that some bars or cables failed (Fiorotto and Rinaldo 1992, Fiorotto and Salandin 2000), and
- a efficient drainage so that pressure fluctuations penetrating through the joints are damped.

5.5   CONCLUSIONS

Ski jumps are frequently used at spillways of high dams. The jet impact location on the plunge pool surface occurs typically distant from the dam toe in order to protect foundations and abutments from dangerous scouring. Further, the jet partially disintegrates and disperses before its impact, so that the specific energy input to the plunge pool and accordingly the scouring potential may be reduced.

The adequate design of a ski jump is essential for its favorable operation, being directly linked to the safety of the dam. A large variety of ski jump types exist, all allowing to influence the jet shape and the impact location on the plunge pool as a function of the spilled discharge, the jet footprint shape, as well as the energy introduced into the plunge pool. An estimation of these parameters is possible, whereas most ski jumps are model tested before construction in order to check their performance (Rao 1982).

Maximum scour will occur if the ski jump operation exceeds a certain duration under a prevailing discharge. If the scour extension and depths endanger the stability of a dam or of the adjacent valley flanks, then protection measures are required. These can include adaptions at the ski jump or an improved operation procedure, or they focus on the plunge pool. There, an increase of the water cushion by providing a tailwater dam or a pre-excavation may be helpful, or a protection of the rock by a well anchored concrete lining. Structural scour protection measures in the plunge pool are normally expensive and have a residual failure risk. Therefore, the selection of the over-all spillway concept is of outstanding importance.

REFERENCES

Berchtold, T., Pfister, M. (2011). Kárahnjúkar Dam spillway, Iceland: Swiss contribution to reduce dynamic plunge pool pressures generated by a high-velocity jet. *Dams in Switzerland*, 315–320, Swiss Committee on Dams, Renens, CH.

Bollaert, E.F.R. (2002). Transient water pressures in joints and formation of rock scour due to high-velocity jet impact. LCH *Communication* 13, EPFL, Lausanne.

Bollaert, E.F.R., Schleiss, A.J. (2003a). Scour of rock due to the impact of plunging high velocity jets 1: A state-of-the-art review. *Journal of Hydraulic Research* 41(5), 451–464.

Bollaert, E.F.R., Schleiss, A.J. (2003b). Scour of rock due to the impact of plunging high velocity jets 2: Experimental results of dynamic pressures at pool bottoms and in one – and two-dimensional closed end rock joints. *Journal of Hydraulic Research* 41(5), 465–480.

Bollaert, E.F.R., Schleiss, A. (2005). Physically Based Model for Evaluation of Rock Scour due to High-Velocity Jet Impact. *Journal of Hydraulic Engineering* 131(3), 153–165.

Bollaert, E.F.R., Duarte, R., Pfister, M., Schleiss, A.J., Mazvidza, D. (2012). Physical and numerical model study investigating plunge pool scour at Kariba Dam. Proc. 24 ICOLD *Congress*. Q94R17, 241–248, Kyoto, J.

Canepa, S., Hager, W.H. (2003). Effect of jet air content on plunge pool scour. *Journal of Hydraulic Engineering* 129(5), 358–365.

Chanson, H. (1994). Drag reduction in open channel flow by aeration and suspended load. *Journal of Hydraulic Research* 32(1), 87–101.

Chanson, H. (1996). *Air bubble entrainment in free-surface turbulent shear flows*. Academic Press, San Diego, US.

Chow, V.T. (1959). *Open-channel hydraulics*. McGraw Hill Book Company INC, New York, US.

Dai. Z., Ning, L., Miao, L. (1988). Some hydraulic problems of slit-type flip buckets. Intl. Symp. *Hydraulics for High Dams*, 1–8, Beijing, CN.

Dhillon, G.S., Sakhuja, V.S., Paul, T.C. (1981). Measures to contain throw of flip bucket jet in installed structures. *Irrigation and Power* 38(3), 237–245.

Di Cristo, C., Iervolino, M., Vacca, A., Zanuttigh, B. (2008). Minimum channel length for roll-wave generation. *Journal of Hydraulic Research* 46(1), 73–79.

Duarte, R. (2014). Influence of air entrainment on rock scour development and block stability in plunge pools. LCH *Communication* 59, EPFL, Lausanne.

Ervine, D.A., Falvey, H.T. (1987). Behavior of turbulent water jets in the atmosphere and in plunge pools. *Proc. Instn Civ. Engrs*, 83(2), Paper 9136, 295–314.

Ervine, D.A., Falvey, H.T., Kahn, A.R. (1995). Turbulent flow structure and air up-take at aerators. *J. Hydropower and Dams* 2(4), 89–96.

Ervine, D.A, Falvey, H.T, and Withers, W. (1997). Pressure fluctuations on plunge pool floors. *Journal of Hydraulic Research* 35(2), 257–279.

Factorovic, N.E. (1952). Gasenije energii pri soudaranii struj potoka. *Gidrotechniceskoe stroitel'stvo* (Hydro-technical construction) 8, 43–44 [in Russian].

Fiorotto, V., Rinaldo, A. (1992). Fluctuating uplift and lining design in spillway stilling basins. *Journal of Hydraulic Engineering* 118(4), 578–596.

Fiorotto, V., Salandin, P. (2000). Design of anchored slabs in spillway stilling basins. *Journal of Hydraulic Engineering* 126(7), 502–512.

Gunko, F.G., Burkov, A.F., Isachenko, N.B., Rubinstein, G.L., Soloviova, A.G., Yuditsky, G.A. (1965). Research on the hydraulic regime and local scour of riverbed below spillways of high-head dams. 9th IAHR *World Conference*, Vol. I, Paper 1.50, Leningrad, RU.

Hager, W.H. (1991). Uniform aerated chute flow. *Journal of Hydraulic Engineering* 117(4), 528–533.

Hager, W.H. (2002). History of roll waves. *L'Acqua* (4), 7–12.

Heller, V., Hager, W.H., Minor, H.-E. (2005). Ski jump hydraulics. *Journal of Hydraulic Engineering* 131(5), 347–355.

Juon, R., Hager, W.H. (2000). Flip bucket without and with deflectors. *Journal of Hydraulic Engineering* 126(11), 837–845.

Kavianpour, M.R., Pourhasan, M.A. (2005). Experimental investigation of pressure fluctuations on the bed of flip bucket spillways. 31st IAHR *World Congress*, 2627–2634, Seoul, Korea.

Kawakami, K. (1973). A study on the computation of horizontal distance of jet issued from ski-jump spillway. *Transactions Japanese Society of Civil Engineers* (5), 37–44 [in Japanese].

Khatsuria, R.M. (2005). *Hydraulics of spillways and energy dissipators*. Dekker, New York.

Lencastre, A. (1985). State of the art on the dimensioning of spillways for dams. *La Houille Blanche* (1), 19–52 [in French].

Lucas, J., Hager, W.H., Boes, R. (2013). Deflector effect on chute flow. *Journal of Hydraulic Engineering* 139(4), 444–449.

Mahzari, M., Arefi, F., Schleiss, A.J. (2002). Dynamic response of the drainage system of a cracked plunge pool liner due to free falling jet impact. *Rock Scour due to falling High-Velocity Jets*, 227–237, Swets & Zeitlinger, Lisse.

Martins, R. (1973). Contribution to the knowledge on the scour action of free jets on rocky river beds. Proc. 11 ICOLD *Congress*. Q41R44, 799–814, Madrid, E.

Mason P.J. (1983). Energy dissipating crest splitters for concrete dams. *Water Power & Dam Construction* (11), 37–40.

Mason, P.J. (1989). Effects of air entrainment on plunge pool scour. *Journal of Hydraulic Engineering* 115(3), 385–399.

Mason P.J. (1993). Practical guidelines for the design of flip buckets and plunge pools. *Water Power & Dam Construction* 45 (9/10), 40–45.

Minor, H.-E. (1987). Erfahrungen mit Schussrinnenbelüftung. *Wasserwirtschaft* 77(6), 292–295 [in German].

Minor, H.-E. (1988). Konstruktive Details zur Vermeidung von Kavitationsschäden. *Mitteilung* 99, D. Vischer ed., 367–378, Laboratory of Hydraulics, Hydrology and Glaciology, ETH Zurich, CH [in German].

Noret, C., Girard J.-C., Munodawafa, M.C., Mazvidza, D.Z. (2013). Kariba dam on Zambezi river: stabilizing the natural plunge pool. *La Houille Blanche* (1), 34–41.

Novak, P., Moffat, A.I.B., Nalluri, C., Narayanan, R. (2007). *Hydraulic structures*. 4th ed. Taylor & Francis, London, UK.

Orlov, V. (1974). Die Bestimmung des Strahlsteigwinkels beim Abfluss über einen Sprung-schanzenüberfall. *Wasserwirtschaft-Wassertechnik* 24 (9), 320–321 [in German].

Pagliara, S., Hager, W.H., Minor, H.-E. (2006). Hydraulics of plane plunge pool scour. *Journal of Hydraulic Engineering* 132(5), 450–461.

Peterka, A.J. (1964). Hydraulic Design of Stilling Basins and Energy Dissipators. USBR *Engineering Monograph* 25, Denver, US.

Pfister, M., Hager, W.H. (2009). Deflector-generated jets. *Journal of Hydraulic Research* 47(4), 466–475.

Pfister, M., Hager, W.H. (2010a). Chute aerators. I: Air transport characteristics. *Journal of Hydraulic Engineering* 136(6), 352–359.

Pfister, M., Hager, W.H. (2010b). Chute aerators. II: Hydraulic design. *Journal of Hydraulic Engineering* 136(6), 360–367.

Pfister, M., Hager, W.H. (2012). Deflector-jets affected by pre-aerated approach flow. *Journal of Hydraulic Research* 50(2), 181–191.

Pfister, M., Hager, W.H., Boes, R. (2014). Trajectories and air flow features of ski jump generated jets. *Journal of Hydraulic Research* 52(3), 336–346.

Rajaratnam, N. (1976). *Turbulent jets*. Elsevier, New York, US.

Ramos, C.M. (1982). Energy dissipation on free jet spillways. Trans. Intl. Symp. *Layout of Dams in Narrow Gorges*. 263–268, Brazilian Committee on Large Dams, Rio de Janeiro, B.

Rao, K.N.S. (1982). Design of energy dissipators for large capacity dams. Trans. Intl. Symp. *Layout of Dams in Narrow Gorges*. Vol. I, 311–318, Brazilian Committee on Large Dams, Rio de Janeiro, B.

Roose, K., Gilg, B. (1973). Comparison of the hydraulic model tests carried out for the ski-jump shaped spillway of the Smokovo and Paliodherli Dams. Proc. 11 ICOLD *Congress*, Q41R37, 671–689, Madrid, E.

Schmocker, L., Pfister, M., Hager, W.H., Minor, H.-E. (2008). Aeration characteristics of ski jump jets. *Journal of Hydraulic Engineering* 134(1), 90–97.

Schleiss, A.J. (2002). Scour evaluation in space and time – the challenge of dam designers. *Rock Scour due to falling High-Velocity Jets*, 3–22, Swets & Zeitlinger, Lisse.

Schleiss, A.J., Annandale, G.W. (2007). State of the art of rock scour technology, Part I. 5th Intl. Conf. *Dam Engineering*, 513–520, Lisbon P.

Straub, L.G., and Anderson, A.G. (1958). Experiments on self-aerated flow in open channels. *Journal of the Hydraulic Division* 84(7), 1–35.

Toombes, L., Chanson, H. (2007). Free-surface aeration and momentum exchange at bottom outlet. *Journal of Hydraulic Research* 45(1), 100–110.

USCE (1977). *Hydraulic Design Criteria*. US Army Corps of Engineers / US Army Engineering Waterways Experiment Station (WES), Vicksburg, Sheets 112–7 and 112–8.

USBR (1974). *Design of small dams*. US Bureau of Reclamation, Denver, US.

Vischer, D.L., Hager, W.H. (1995). *Energy Dissipators*. IAHR Hydraulic Structures Design Manual 9, A.A. Balkema, Rotterdam, NL.

Vischer, D.L., Hager, W.H. (1998). *Dam Hydraulics*. John Wiley & Sons, Chichester, UK.

Wood, I.R. (1991). Free surface air entrainment on spillways. IAHR *Hydraulic Structures Design Manual* 4, 55–84, A.A. Balkema, Rotterdam, NL.

Zhenlin, D., Lizhong, N., Longde, M. (1988). Some hydraulic problems of slit-type flip buckets. Intl Symp. *Hydraulics for High Dams*, 287–294, Beijing, CN.

Chapter 6

Impact dissipators

B.P. Tullis & R.D. Bradshaw
Department of Civil and Environmental Engineering, Utah State
University, Logan, Utah, USA

ABSTRACT

Flow impacting an obstruction plays a role in many different types of energy dissipators. In particular, hanging baffle dissipators and baffled chutes principally use impact dissipation to achieve acceptable downstream flow conditions. This chapter provides general guidance on the use and design of these two dissipators, as well as examples of design variations employed in practice to adapt for specific application needs.

6.1 INTRODUCTION

Reduction of kinetic energy in flows through principles of impact is a feature of many types of energy dissipation structures. Impact dissipation is characterized by the impact of the flow on an obstruction placed in its path, resulting in a loss of momentum due to the forced change in flow direction. The turbulence and mixing created by the sudden redirection further contributes to the overall energy dissipation. The nature of this method of energy dissipation typically limits its application to relatively low velocity flows in order to avoid damage to the structure through cavitation, abrasion, and/or the dynamic forces involved.

The term "impact dissipator" is used somewhat ambiguously in the literature. Therefore, this chapter will focus mainly on those dissipation structures that could be termed "true impact dissipators" (i.e., function effectively using only the principles previously described). These dissipators do not necessarily rely on the formation of a hydraulic jump nor require the presence of tailwater. This chapter will examine two such dissipators in detail: the hanging baffle dissipator and the baffled chute. A number of alternative structures that employ impact dissipation elements are briefly discussed in Section 6.4.

6.2 HANGING BAFFLE

6.2.1 Overview

Impact dissipators that feature a hanging baffle wall with an inverted 'L' shaped cross-section (see Fig. 6.1) are known by many names, including: hanging baffle dissipators, baffled outlets, baffle boxes, outlet stilling basins, impact basins, USBR Type-VI

Figure 6.1 Examples of hanging baffle dissipators: (A) Julius Park Reservoir (UT, USA) (courtesy of Bret Dixon, Utah Division of Water Rights Dam Safety Program), (B) Lost Lake (UT, USA).

basins, or simply impact-type dissipators. The first designs of this type were developed by the United States Bureau of Reclamation (USBR) in response to the energy dissipation needs of the Missouri River Basin Project (now called the Pick-Sloan Missouri Basin Program). Canals in this project required a relatively compact energy dissipation structure that could effectively reduce excess kinetic energy regardless of tailwater conditions. The hanging baffle feature was developed as a means of converting a variety of incoming discharge, depth, and velocity conditions into a common exit flow pattern (USBR, 1955). As this irrigation project would require the construction of more than 50 such structures, the USBR developed standardized design guidelines for the hanging baffle dissipator in 1955 (revised by Beichley, 1978). Hanging baffle dissipators have since become a standard design for closed conduits to open channel flow transitions where site requirements prohibit the use of a free-discharging jet.

Dissipation via a hanging baffle structure is achieved mainly through direct impact, as well as the resulting flow turbulence and mixing. Flow enters the concrete dissipator and strikes the vertical face of a slab suspended from the sidewalls. This hanging baffle absorbs a portion of the energy on impact and redirects the flow in all directions. As water builds up behind the hanging baffle, the basin floor, sidewalls, and baffle overhang contain the spray and redirect the flow back onto itself, creating turbulent eddies that further contribute to the reduction of kinetic energy. The water then exits through a gap between the floor and the bottom of the baffle, where it passes through a short stilling basin and finally over an end sill to the receiving water body.

6.2.2 Design guidelines

The USBR Type VI Stilling Basin design found in *USBR Research Report No. 24: Hydraulic Design of Stilling Basin for Pipe or Channel Outlets* (Beichley, 1978) is the most current USBR publication on hanging baffle dissipator designs and provides the basis for most hanging baffle design methods. The USBR recommends that their design standards be applied only within the following limits:

–   Froude Numbers ranging from 1 to 10 (Beichley, 1978)
–   Maximum inlet velocity of 15 m/s (50 ft/s) (Beichley, 1978)
–   Maximum design discharge of 11 m³/s (400 ft³/s) (USBR, 1987)

These limits represent the extent of the model testing performed by the USBR and are the point at which it was feared that damage due to cavitation, impact forces, or vibration could occur. In cases where a discharge larger than 11 m³/s (400 ft³/s) must be passed, the flow may be split and directed into multiple parallel units, often constructed with a common wall between them (as shown in Fig. 6.2) (Peterka, 1978).

The USBR hanging baffle dissipator geometric dimensions are all a function of the dissipation basin width, W. The magnitude of W is determined using the Froude Number (*Fr*), just upstream of the basin entrance, and Equation (6.1):

$$\frac{W}{d*} = 2.868\left(\mathrm{Fr}^{0.5664}\right) \tag{6.1}$$

Figure 6.2 Two parallel hanging baffle dissipators constructed to accommodate discharge capacities beyond the standard limits at Brundage Dam (ID, USA) (courtesy of Idaho Department of Water Resources Dam Safety Program).

where d^* is the characteristic length dimension for the Froude number; in this application, d^* is the square root of the flow cross-sectional area in the upstream pipe (i.e., d^* represents the equivalent flow depth in a square conduit of equivalent flow cross-sectional area to the actual conduit flow) (Beichley, 1978). The resulting design curve is presented in Figure 6.3. A diagram showing the USBR guidelines for sizing the dissipator is given in Figure 6.4.

When sizing a hanging baffle dissipation basin, the following guidelines should be observed:

–   Oversizing the basin should be avoided as this may cause the incoming jet to impinge only partially upon the hanging baffle, resulting in reduced energy dissipation.
–   An undersized hanging baffle basin can result in excessive splashing and spray escaping the basin, increasing the risk of embankment erosion (see Fig. 6.5).
–   Dissipator design, including placement of steel reinforcement in the concrete, must be structurally sound and capable of resisting vibration, basin movement, and other adverse dynamic force effects (USBR, 1987; Young, 1978).

The approach pipeline should be considered in the design phase as well. Beichley (1978) also provides guidelines for alternative downsloping conduit angles entering

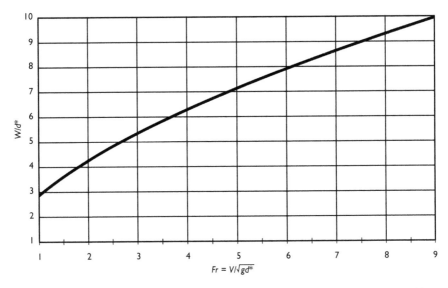

Figure 6.3 Sizing curve for hanging baffle dissipator, where W is the minimum width of the dissipator basin and d^* is the significant depth used in calculating the Froude Number (after Beichley, 1978).

the dissipator (horizontal is typical), to ensure proper flow impact on the hanging baffle.

Figure 6.4 shows alternate wing wall and end sill options for the basin outlet. The angled wing walls expand the discharge flow area at the downstream end of the basin to reduce the exit velocity. The wing walls are recommended but not required per the USBR design method. Verma & Goel (2000) examined the performance of different end sill shapes and found that both an angled sill (such as the alternate end sill shown in Fig. 6.4) and a sill with a quarter-rounded leading edge produced better scour characteristics than a square-edged sill. A small notch or drain should be included in the end sill to facilitate basin drainage during periods of low or absent flow (Thompson & Kilgore, 2006; UDFCD, 2008). The trapezoidal notches in the bottom of the hanging baffle, shown in Figure 6.4, are optional elements that improve the self-cleaning nature of the basin during normal operation (discussed further in Section 2.3).

Riprap should be placed downstream of the end sill as shown in Figure 6.4 to avoid erosion of the downstream channel banks and invert. While any reliable method (e.g., Peterka, 1978; Thompson & Kilgore 2006; or USACE, 1970) may be used to determine stone size and depth of placement, Beichley provides a conservative estimate of basin width to stone diameter of 20:1, placed to a depth and elevation equal to the height and elevation of the end sill, respectively. Additional stability and protection from undermining can be obtained by adding cutoff walls at the downstream end of the dissipator (Thompson & Kilgore, 2006).

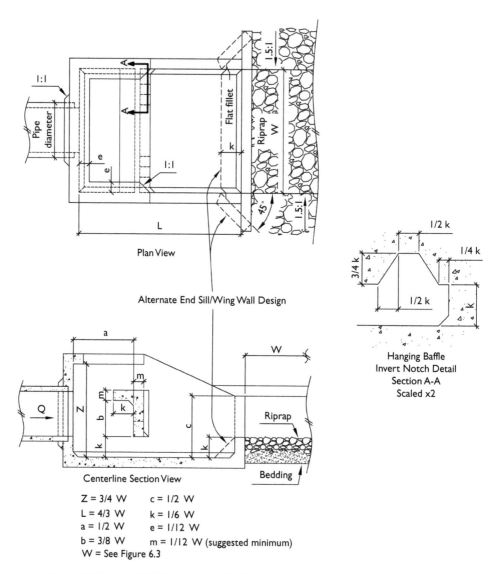

Plan View

Alternate End Sill/Wing Wall Design

Hanging Baffle
Invert Notch Detail
Section A-A
Scaled x2

Centerline Section View

Z = 3/4 W   c = 1/2 W
L = 4/3 W   k = 1/6 W
a = 1/2 W   e = 1/12 W
b = 3/8 W   m = 1/12 W (suggested minimum)
W = See Figure 6.3

Figure 6.4 Standard USBR hanging baffle dissipator dimensions (after Beichley, 1978).

6.2.3 Advantages & limitations

As the majority of energy dissipation occurs before the flow passes beneath the hanging baffle, very little downstream stilling basin length is typically required. Hanging baffle dissipator basins are therefore typically much shorter than most stilling basins that utilize a hydraulic jump, which often translates to economic advantages (Young, 1978; Beichley, 1978). In addition to improved economy, the hanging baffle dissipator has also been shown to be more efficient than a basic hydraulic jump-type

Figure 6.5 Intense splashing due to insufficient space between the conduit outlet and hanging baffle [Figure 54 in *Outlet Works Energy Dissipators* published by the United States Federal Emergency Management Agency (FEMA) (FEMA, 2010)].

dissipator in that it provides a greater reduction in energy at a given Froude Number, as shown in Figure 6.6.

By not utilizing a hydraulic jump as the primary means of energy dissipation, this dissipator also has the advantage of having no minimum tailwater depth requirement, per the USBR design method (Beichley, 1978). The ability to function well in the absence of tailwater has made hanging baffle dissipators a popular choice for applications where discharge may be intermittent or may fluctuate suddenly, such as with storm sewer outlets. While no tailwater is required for satisfactory performance, the presence of tailwater does produce a smoother downstream surface profile and lower exit velocities (Beichley, 1978). Rice & Kadavy (1991) and Blaisdell (1992) suggest that tailwater enhancement of dissipator performance only occurs for tailwater-depth-to-pipe-diameter ratios > 0.65. The performance benefits produced by the presence of a consistent tailwater are not sufficient to warrant a reduction in the length of the dissipation basin; however, the corresponding reduction in velocity may permit a reduction in the longitudinal extent and size of the riprap downstream. According to Beichley (1978), the tailwater depth should not exceed $k + b/2$ (see Fig. 6.4) above the basin invert ($k + b/2 \approx$ hanging baffle midline elevation).

For cases in which the dissipator experiences long periods of non-use, wind-blown material may tend to accumulate in the basin. At installations where this is a concern, trapezoidal notches such as those shown in Figures 6.1B and 6.4 may be added to the

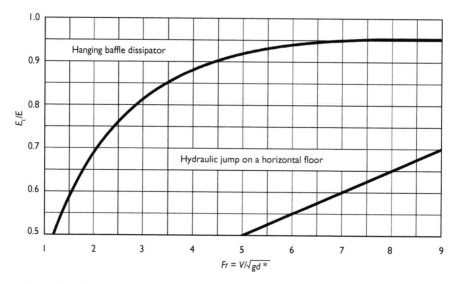

Figure 6.6 Efficiency of a hanging baffle dissipator versus that of a basic hydraulic jump basin, where E_L/E is the ratio of energy loss in the basin to the flow energy at the entrance (after Beichley, 1978).

hanging baffle invert to improve the self-cleaning nature of the basin. These notches allow the formation of concentrated jets that help transport the sediment from the dissipator (Young, 1978; Beichley, 1978). While this is an effective means of preventing clogging due to sediment buildup, there is some evidence that inclusion of the bottom notches may increase scour potential downstream of the basin (Verma & Goel, 2000), suggesting that the notches should be omitted from the design where sedimentation is not a concern.

Larger debris may tend to clog the opening between the basin floor and the hanging baffle (tumble weeds can be particularly problematic). As removal of debris from behind the baffle is inconvenient and often difficult, it is advised that fencing, screens, or grates be placed above the dissipator if the presence of larger debris introduced by wind, vandalism, or other methods is anticipated (see Fig. 6.7; also the grating placed over the basin in Fig. 6.2). However, as a safety measure, the standard design was developed to accommodate the entire discharge passing over the top of the hanging baffle for a limited time, albeit with less efficiency, in the event that the opening below the hanging baffle becomes blocked (Beichley, 1978). Fencing also improves public safety.

6.2.4 Variations & further research

Many institutions have taken the standard USBR design and altered it according to their experience and specific needs. The design guidelines published by the Urban Drainage & Flood Control District (UDFCD) in Colorado, USA (2008) are one example. The UDFCD needed to divide storm water discharges into multiple, smaller

Figure 6.7 Example of a hanging baffle dissipator with fencing and signage to help prevent debris from entering the basin.

diameter pipes to accommodate low headroom clearances at street crossings. UDFCD adapted the USBR design to this situation by developing modifications that allow for the discharge of multiple conduits into the same dissipator (Urbonas, 2013).

The hanging baffle dissipators at the Pima Mine Road groundwater recharge basins in Arizona, USA (Central Arizona Project) feature steel plate armoring on the impact side of the hanging baffles to provide protection against the abrasive nature of the discharge commonly observed at the site. The flow discharging into these dissipators issues from jet flow gate valves that are housed in open vaults immediately upstream of the hanging baffle dissipators. Based on a valve manufacturer suggestion that the valve should not operate submerged, the dissipator design was altered to angle the hanging baffle away from the vault in order to reduce submergence potential (Jackson et al. 2013), as shown in Figure 6.8.

Other designs have included placement of a pillar beneath the hanging baffle for structural support (Thompson & Kilgore, 2006; TDOT, 2010). This modification has also been implemented in order to keep low flows from jetting below the baffle unchecked (Lemons, 1975). Baffle blocks have also been added to the basin invert (see Fig. 6.9). While this provides additional dissipation for a relatively small increase in construction costs, it is unclear as to whether these baffle blocks have an appreciable impact on the overall basin performance. While the majority of the variations discussed lack laboratory validation, they do provide a starting point for future investigation and individual adaptation.

Figure 6.8 Modified hanging baffle dissipator in Arizona, USA. The hanging baffle is slightly angled away from the incoming jet. The bolts that anchor the steel plating can be seen. The increased drop height of the dissipator was a site constraint.

Recent hanging baffle dissipator research has focused primarily on further reduction of downstream scour potential. Research by Tiwari (2013) suggests that better scour performance can be obtained by moving the hanging baffle farther downstream (increasing dimension a in Fig. 6.4) and/or by increasing the baffle height and overhang length (dimensions b and k, respectively, in Fig. 6.4).

Verma & Goel (2000) experimented with adding additional appurtenances to the basin in order to reduce scour, such as an intermediate sill below the hanging baffle to create a roller in the basin and a splitter block aligned with the incoming flow to spread it more evenly across the basin width. They found that these features could

Figure 6.9 A construction photo from the Pima Mine Road project in Arizona, USA. This hanging baffle dissipator features baffle blocks downstream of the hanging baffle for additional energy dissipation.

lead to major reductions in the scour of an unprotected streambed downstream, suggesting the possibility of shorter basin lengths. Goel & Verma (2001) and Verma & Goel (2003) further evaluated the effects of adding appurtenances; however, these two studies featured outlet pipe invert elevations consistent with that of the basin floor, which calls into question whether the hanging baffle is effectively utilized.

The USBR produced a small-application hanging baffle dissipator design for the United States Forest Service to be used at trail and roadway underdrains that can be easily fabricated from sheet metal (Colgate, 1971). The USBR (1958) also conducted model tests for variations on the standard design for specific sites that required entrance velocities of up to 32 m/s (106 ft/s) and a steep supply pipe angle.

Additional research may be warranted for many of the design variations previously described. The maximum velocity and discharge limits [15 m/s (50 ft/s) and 11 m³/s (400 ft³/s)] associated with the USBR design and propagated throughout the subsequent literature, which were intended to minimize negative cavitation and vibration effects, may be overly conservative based on the limited amount of data supporting those limits. These limits appear to correspond to maximum field-application velocities and discharges observed at the time rather than on hydraulic testing.

Model tests performed by the USBR (1958) for a case-specific application examined the performance of a hanging baffle dissipator downstream of a gate valve. The design was similar to that presented later by Beichley (1978), but operated at a velocity of 24 m/s (80 ft/s) and a discharge of 0.28 m³/s (10 ft³/s). They found the energy dissipation performance under these conditions to be acceptable; however, additional modifications were required to reduce the intense splash that resulted. No mention is made of adverse vibrational or cavitational side effects, suggesting that this type of dissipation structure may be able to tolerate higher velocities, especially if the structure seldom experiences the design discharge.

6.3 BAFFLED CHUTE

6.3.1 Overview

The baffle chute is one of the most time-tested structures for dissipating energy at canal grade changes (see Fig. 6.10) in a wide variety of irrigation applications (Peterka, 1978). Alternative names for the baffled chute include baffled drop, baffled apron, or USBR Type IX dissipator. The baffled chute dissipator features staggered rows of baffles that reduce the acceleration of the flow, through impact and redirection, as it proceeds from row to row. The baffles restrict flow velocities to acceptable levels

Figure 6.10 Example of a baffled chute canal drop structure near Willard, Utah, USA.

along the chute length. The baffled chute requires no downstream stilling basin and its energy dissipation effectiveness is unit-discharge dependent but independent of the total change in water surface elevation across the structure. Alternative applications for baffled chutes include closed-conduit to open channel flow transitions (Peterka, 1978; FEMA, 2010), similar to the hanging baffle. In addition to their successful use in agricultural settings, the baffled chute has also proven effective for energy dissipation in urban settings and for highway drainage (Rhone, 1977; Thompson & Kilgore, 2006). Although perhaps more commonly used for canal grade change application, baffled chutes have also been adapted to spillway application (Rhone, 1977) as shown in Figure 6.11.

6.3.2 Design guidelines

The USBR first published standardized design guidelines for baffled chute dissipators in 1963 as part of *Hydraulic Design of Stilling Basins and Energy Dissipators* (Peterka, 1978) and later in *Design of Small Canal Structures* (Hayes, 1978). Both of these resources were developed using physical model study data and observing the performance of many existing prototype structures.

The USBR's design parameters for smaller-scale canal baffled chutes (as opposed to larger spillway designs) are based on design unit discharge, $q_{des} = Q/W$, where Q is

Figure 6.11 Example of wide baffled chute emergency spillway at DMAD Reservoir near Delta, Utah, USA [width = 61 m (200 ft), design unit discharge = 7.2 m²/s (78 ft²/s)].

the total design discharge and W the dissipator chute width. The value of W may be influenced by site conditions but should be selected so that q_{des} falls within the acceptable limits presented here. The chute and baffle block dimensions are all a function of q_{des}. The model studies upon which the USBR based their recommendations showed that the lower the value of q, the less severe the flow conditions at the outlet of the chute and the potential for scour. In connection with these observations, the USBR provides the following general guidelines:

- $q_{des} \leq 5.6$ m²/s (60 ft²/s) is the upper limit for recommended unit discharges when some downstream degradation and scour pool formation at the base of the chute can be tolerated.
- $q_{des} \approx 3.3$ m²/s (35 ft²/s) results in measurable but less severe downstream scour conditions.
- $q_{des} \leq 1.9$ m²/s (20 ft²/s) produces mild flow conditions downstream and minimal scour potential (Peterka, 1978).

It is common for baffled chutes to experience some amount of scour at the downstream toe. The USBR design assumes that some scour will occur by specifying that the chute be extended below the final grade of the receiving channel (see Fig. 6.12), and then backfilled with soil or loose rock. Over time, a stable scour pool will form. Coupled

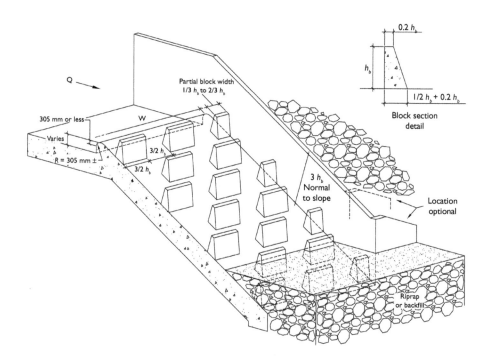

Figure 6.12 USBR dimensions for a baffled chute (cut down the longitudinal centerline). In the standard USBR design, the first row of blocks should be located no more than 305 mm (12 in) below the crest elevation and should include the partial blocks connected to the sidewalls (after Peterka, 1978).

with the additional blocks uncovered by the scour, this pool further aids in reduction of any excessive velocities at the end of the chute. The chute should be extended far enough below grade that it prevents undermining of the structure from any anticipated scour or downstream degradation at the design flow. A minimum of one additional row of blocks buried at the base of the chute is recommended (Peterka, 1978).

Observation of baffled chute prototypes revealed that some of these structures, designed for the maximum recommended design unit discharge [5.6 m²/s (60 ft²/s)], had proven capable of safely passing approximately 2-times that amount for "short" periods of time. This led to an investigation of whether the USBR design guidelines could accommodate appreciably larger unit discharges for use in larger spillway applications. Rhone (1977) conducted hydraulic model tests on baffled chutes designed according to the guidelines presented by Peterka (1978) and Hayes (1978) and extrapolated for q_{des} values of up to 28 m²/s (300 ft²/s). Rhone described the resulting flow conditions as being "essentially the same as had been found with smaller unit discharges" and that erosion at the base of the chute was "moderate" for all values of q evaluated. It is assumed that the description offered by Rhone of moderate erosion and similar flow characteristics to those of the canal-scale design indicate that the same trend of greater scour at greater unit discharges holds true. Formation of a scour pool at the base of the chute and some amount of downstream degradation should then be expected at these higher unit discharges.

Rhone suggests that, while any $q_{des} > 5.6$ m²/s (60 ft²/s) should be validated through hydraulic model testing, economic factors and constructability may be the primary limit regarding the maximum unit discharge that can be designed for. This assumes that the chute extends far enough below existing grade to accommodate the scour associated with these flows. A hydraulic study conducted by George (1979) with $q = 11.7$ m²/s (126 ft²/s) uncovered multiple rows of buried baffle blocks. Further model studies have found that for larger design unit discharges ($q_{des} > 5.6$ m²/s (60 ft²/s)), the chute may be dimensioned based on a q_{des} of two-thirds the actual value, so long as the height of the sidewalls is increased to accommodate the full design unit discharge (USBR, 1987). The designer may find that it is more economical to widen the chute, thereby decreasing q_{des}, along with the corresponding baffle block dimensions and chute wall height requirements.

Once q_{des} is determined, the critical depth, d_c, and critical velocity, V_c, for rectangular channels are determined using Equations (6.2) and (6.3).

$$d_c = \sqrt[3]{\frac{q^2}{g}} \tag{6.2}$$

$$V_c = \sqrt[3]{gq} \tag{6.3}$$

A baffle block height, h_b, of $0.8d_c$ to $0.9d_c$ is recommended (see design curves shown in Fig. 6.13). The remaining chute and baffle block dimensions are determined from Figure 6.12. Apart from the height, the block dimensions are flexible and may vary, provided that structural stability is maintained. Hayes (1978) suggests that the baffle blocks may be precast in order to reduce expenses. A chute slope, S, of 2:1 (horizontal:vertical) or flatter is recommended. The longitudinal spacing of the staggered baffle block rows

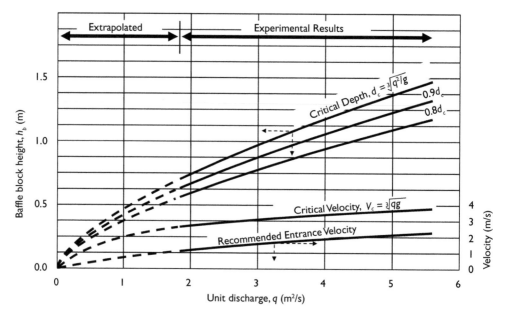

Figure 6.13 Design curves for selecting baffle block height and entrance velocity (after USBR, 1987). Rhone (1977) suggests that these relationships may be extrapolated for higher values of q.

should be equal to h_b/S, where S is in decimal form. The flow will typically pass through multiple rows of baffles before a fully established flow pattern is achieved. When $q \leq q_{des}$, the flow pattern in the chute is typically fully developed within the first four rows of baffle blocks; the remaining rows arresting any further acceleration as the flow proceeds down the chute, as illustrated in Figure 6.14 (USBR, 1987).

Designing the structure to provide the proper approach flow characteristics is critical to the effectiveness of the design. As Peterka (1978) describes, "The baffled apron is not a device to reduce the velocity of the incoming flow; rather, it is intended only to prevent excessive acceleration of the flow passing down the chute." Flows that approach critical velocity at the chute entrance can strike the first row of baffles with such force that the flow is deflected up and over subsequent rows and potentially out of the chute. Therefore, the approach velocity entering the chute should be well below critical; the USBR (1987) recommends an entrance velocity of at least 1.5 m/s (5 ft/s) less than V_c, as shown in Figure 6.13. In order to establish a well-distributed, subcritical entrance flow regime, it may be necessary to incorporate a small stilling basin at the head of the chute and/or to offset the chute crest from the approach channel floor (see Fig. 6.12). However, this should be avoided when possible as it can result in collection of sediment or other debris at the head of the chute (UDFCD, 2008).

Chute sidewall heights should be at least $3\,h_b$ to contain the flow (normal to chute floor). Designing sidewalls sufficiently tall to contain all of the spray produced may be economically impractical for typical applications. Subsequently, the embankment adjacent to either side of the chute should be properly protected from spray-induced

Figure 6.14 The flow regime in this baffled chute appears to be fully established after the fourth row of baffle blocks.

erosion with vegetation, riprap, or a combination of the two. Baffled chutes should also be protected against uplift, sliding, and other dynamic forces via cutoff walls and/or protective drains. Wing walls may also be included both upstream and downstream of the structure to protect the channel embankments from erosion (Hayes, 1978). Figure 6.12 shows different downstream location options for 90° wing walls. Moving the wing walls upstream, which results in a higher top-of-wall elevation, may be preferable to protect the embankment adjacent to the outlet if significant tail water elevations are anticipated.

6.3.3 Advantages & limitations

One of the key characteristics of the baffled chute is that it is effective regardless of tailwater conditions (although presence of tailwater does improve scour performance). Additionally, once the flow has passed enough rows of baffle blocks for full flow development (no additional net flow acceleration), the chute can be extended to any length by continuing the staggered rows of blocks, without impacting its effectiveness (assuming the slope and baffle block guidelines presented previously are followed) (FEMA, 2010; USBR, 1987).

As scour at baffled chute outlets is dependent on q, the guidelines presented previously should be used to select a value of q_{des} that is likely to produce scour that falls within the acceptable limits for a particular site. The practice of extending the chute below the downstream channel grade has the distinct benefit of not only protecting the structure from downstream degradation and scour (whether anticipated or due to higher than expected maximum q values), but of also continuing to provide effective energy dissipation under these conditions.

As can be seen in Figure 6.15, the baffled chute produces fully aerated flow when at least 5–7 rows are present (FEMA, 2010). As a side benefit, Rhone (1977) observed that in addition to aeration, the turbulence created by baffled chutes is also capable of significantly reducing levels of dissolved nitrogen gas in the flow; however, the mechanism behind this off gassing was not discussed. This suggests that the baffled chute may be an exceptional dissipator choice where such environmental concerns are present.

There is some disagreement in the literature regarding the degree to which accumulation of debris is a problem in baffled chutes. In many of the prototypes observed by Peterka (1978), debris was mainly deposited on the chute in the final stages of a flow event (falling leg of the flood flow hydrograph), and then carried away by the

Figure 6.15 Example of fully-aerated flow in a baffled chute.

rising hydrograph leg of the next event (see also FEMA, 2010). However, concern has been expressed regarding the fact that debris may clog the baffles, causing the chute flow to overtop the sidewalls (UDFCD, 2008; Hayes, 1978). For situations in which heavy debris may be present, it is recommended that appropriate measures be taken to prevent this debris from entering the chute (e.g., trash racks).

6.3.4 Variations & further research

Variations between baffled chute designs focus mainly on differences in the baffle block and chute crest details. Once the baffle block height has been established, the USBR design guidelines allow for some flexibility in the remaining dimensions. A common variation tested by the USBR was whether the upstream block face should be oriented vertically or normal to the chute floor; little difference was observed in the performance of the two variations and the normal face was selected as the standard (Peterka, 1978).

As the standard block shape has been extensively evaluated by the USBR and exhibited effectiveness over the past 50+ years, few have experimented with changing its basic design. One exception would be a uniquely shaped baffle developed by the United States Army Corps of Engineers (USACE), shown in Figure 6.16. The USACE (1990) reported that rows of these blocks (row and block spacing also differs from the USBR guidelines) performed well, with good aeration performance for $q \leq 17$ m²/s (180 ft²/s) and adequate energy dissipation for $q \leq 84$ m²/s (900 ft²/s).

Alternate chute crest designs were developed in response to a need to prevent backwater effects upstream of the chute. While the USBR guidelines specify that the first row of blocks should be as close to the crest as possible in order to prevent flow acceleration at the top of the chute, backwater effects in the approach flow are created because the top elevation of these blocks exceeds the crest elevation, potentially creating a new control point.

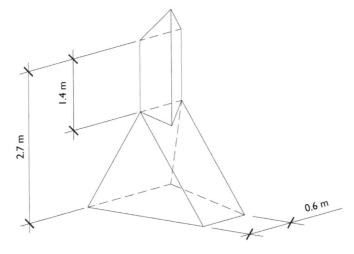

Figure 6.16 Alternate baffle block developed by the United States Army Corps of Engineers (after USACE, 1990).

As this backwater effect may be undesirable for some site-specific applications, two main alternate chute crest designs were developed: the Fujimoto entrance and the Type XVI entrance. Consistent with the USBR standard design, these alternative crest transition geometries require subcritical approach flow but also help reduce backwater effects upstream. The Fujimoto entrance, shown in Figure 6.17, has been used successfully for chutes with larger unit discharges [$q > 9.3$ m²/s (100 ft²/s)]. The rounded crest is replaced with a horizontal, dentated broad crested weir. The regular baffle blocks then begin approximately where the third row would start per the standard USBR design, although the optimal location should be determined via a hydraulic model study (Rhone, 1977; USBR, 1987).

Less common than the Fujimoto entrance, the Type XVI entrance consists of a crest that is triangular in cross section with a vertical upstream face. Triangular sidewall contractions are added at the distal ends of the crest, oriented normal to the sidewall and chute floor, creating a cipolletti-type contracted weir condition. The first row of baffles of the USBR design is also removed, as shown in Figure 6.18. Rhone (1977) claims that the arrangement does not cause excessive acceleration before first impact. Both entrance types aid in preventing backwater effects due to debris clogging;

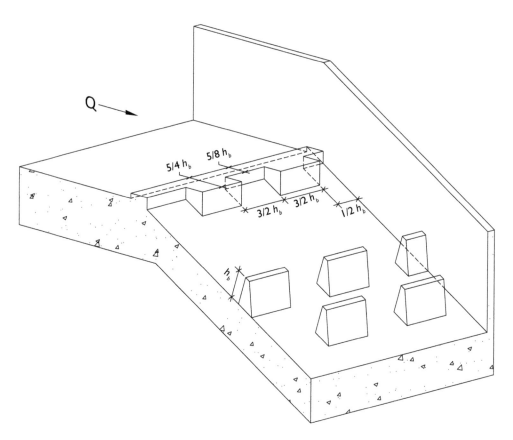

Figure 6.17 Fujimoto entrance (after Rhone, 1977).

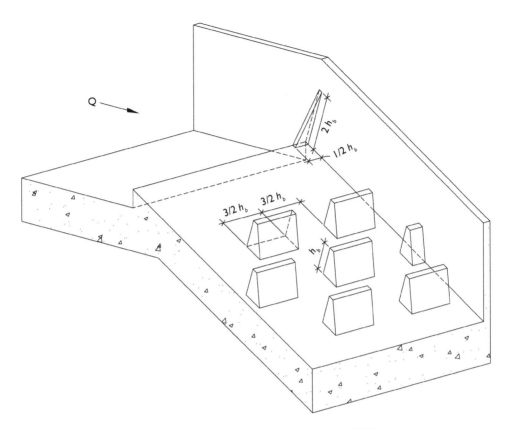

Figure 6.18 Type XVI entrance (after Rhone, 1977).

however, debris accumulation farther down the chute may still cause sidewall over-topping. Some maintenance may still be required to keep the chute free of debris, though this can be reduced by incorporating trash racks or other similar measures.

For structures with large elevation changes, the 2:1 maximum baffled chute slope could potentially be a limitation due to the relatively large longitudinal footprint required to accommodate the baffled chute. Higgins (1981) suggested the possibility of incorporating a 90° horizontal turn in the chute in order to extend this dissipator design to a wider variety of site applications. Higgins found that when guide vanes are incorporated into the horizontal bend, the level of scour potential at the down-stream toe remained consistent with the traditional design. However, the application of Higgins' results is somewhat limited in that all the tests were run at $q = 3.7$ m²/s (40 ft²/s). Additional riprap was also required for protection of the embankment at the turn, as some of the typical splash in the chute was carried beyond the sidewalls.

While the magnitude and/or extent of scour is often a key parameter in evaluat-ing the effectiveness of various design changes, little has been done to quantify the relationship between q and expected scour. The studies referenced in this section use scour data largely as a means of making relative comparisons between design variations or flow conditions (e.g., using "moderate," "severe," or "excessive" without quantifying what these differences mean from a practical standpoint). The functionality of the baffled chute design guidelines could be improved by further research efforts to better define the impacts of q on scour. This would provide better guidance for selecting more appropriate site-specific q_{des} values based on allowable scour. Additional research would help validate the design variations presented in this section, as well as to investigate the mechanisms behind the observed nitrogen off gassing and to quantify the effects of q on the severity of splash outside of the chute.

6.4 ADDITIONAL USES OF IMPACT DISSIPATION

While the hanging baffle and baffled chute dissipators rely solely on impact dissipation for their effectiveness, there are many dissipators that utilize principles of flow impact to improve their performance. This section briefly discusses dissipator designs not strictly categorized as "impact dissipators," but that include impact dissipation in their design.

6.4.1 Hydraulic jump

Hydraulic jump formation is perhaps the most common method of reducing kinetic energy in water flow. A simple, horizontal stilling basin may require a significant stream-wise length to facilitate hydraulic jump formation for larger unit discharges. By adding impact elements such as chute blocks, baffle blocks and end sills, other hydraulic jump basin designs (e.g. USBR Type-II, Type-III, and Type-IV standard design basins or the St. Anthony Falls stilling basin) have been effective at reducing hydraulic jump stilling basin lengths by up to 60% (see chapter 4 for more details on hydraulic jump energy dissipation basins).

A variety of lesser known basin designs, such as the contra costa basin, "standard" baffle dissipator, hook dissipator basin, and others are discussed by Baston (2000), FEMA (2010), and Thompson & Kilgore (2006). Note that many of these alternate designs have minimal design support data and/or field application experience.

6.4.2 Stilling wells

A vertical shaft stilling well (see Fig. 6.19) can also be used to transition from closed conduit to free surface flow. The floor and/or walls of the well alter flow momentum as flow enters the well near the invert; kinetic energy is dissipated by turbulent mixing as it flows vertically through the well toward the outlet. Optional sidewall baffles can further increase energy dissipation (see Fig. 6.19). Protective plating on the well bottom surfaces is often required to prevent cavitation damage (USBR, 1987).

6.4.3 Fixed cone valve with baffled hood

The fixed-cone valve (also known as a Howell-Bunger® valve), often used as a free-discharging valve, distributes discharge into a conical spray. The conical jet is

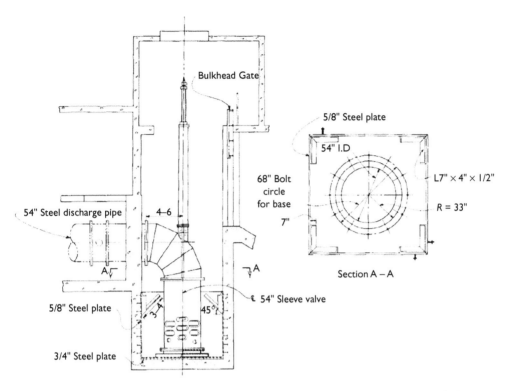

Figure 6.19 Example of a vertical stilling well with a sleeve valve and protective steel plating [Figure 10–17 from *Design of Small Dams* (USBR, 1987)].

very effective at increasing jet air entrainment and significantly reducing the unit discharge entering the plunge pool by distributing the flow over a larger area. For applications where a sufficiently large downstream area to facilitate the conical jet is not available, a hood is often added to the fixed cone valve in order to contain and redirect the spray into a straight, hollow jet which then enters the plunge pool (see Fig. 6.20). Baffles or "teeth" can be added to the fixed-cone valve hood as an impact element (see Fig. 6.21) to increase the energy dissipation effectiveness and to reduce the extent of the spray area, thus reducing the required plunge pool size.

Johnson & Dham (2006) reported a significant increase in energy dissipation through the use of baffles in a fixed cone valve hood. Their solution features three rows of staggered baffles mounted inside a short hood (similar to that shown in Fig. 6.21). This "baffled hood" effectively reduces the flow exit velocity and jet trajectory length, facilitating a reduction in plunge pool size, as shown in Figure 6.22. For this particular application, the baffled hood successfully dissipated up to 92% of the flow energy entering the fixed cone valve. While work remains to be done in developing this concept, the design has already been implemented successfully in the field (Johnson et al. 2001).

Figure 6.20 Discharge from a fixed cone valve equipped with a hood (no baffles) (Courtesy of Michael Johnson, Utah Water Research Laboratory).

Figure 6.21 Example of a baffled fixed cone valve hood with staggered rows of impact baffles (Courtesy of Michael Johnson, Utah Water Research Laboratory).

Figure 6.22 Discharge from a fixed cone valve equipped with a baffled hood; notice the difference in intensity compared to the hooded fixed cone valve shown in Figure 6.20 (Courtesy of Michael Johnson, Utah Water Research Laboratory).

ACKNOWLEDGEMENTS

Contributions to the development of this chapter were made by the following: Philip Thompson (Hydraulics Specialist); Patrick Dent (Central Arizona Project); Lisa Jackson, Pablo Gonzalez-Quesada, and Ahmed Hussein (Black & Veatch); Jim Walker and Megan Greathouse (Sevier River Water Users Association); Kent Welling (City of Los Angeles Hydraulics Research Laboratory); Mike Hand (Wyoming Surface Water Division – Safety of Dams Program); Michael Johnson (Utah Water Research Laboratory); Bret Dixon and Everett Taylor (Utah Division of Water Rights – Dam Safety Program); Daryl Devey (Central Utah Water Conservancy District); John Falk (Idaho Dept. of Water Resources – Dam Safety Program); Jerry Chaney (Utah Department of Transportation), Ben Urbonas (Urban Watershed Research Institute), Tony Wahl (U.S. Bureau of Reclamation).

REFERENCES

Baston, C.R. (2000). *Preliminary Investigation of Culvert Outlet Baffle Block Geometry and Energy Dissipation*. [Online] Morgantown, WV: West Virginia University. Available from: http://wvuscholar.wvu.edu/ [Accessed on 10 July 2013].

Beichley, G.L. (1978). *Hydraulic Design of Stilling Basin for Pipe or Channel Outlets*. [Online] United States Bureau of Reclamation. Research report number 24. Available from: http://www.usbr.gov/pmts/hydraulics_lab/reportsdb/reportsearchall.cfm [Accessed 15 May 2013].

Blaisdell, F. (1992). Discussion of "HGL Elevation at Pipe Exit of USBR Type VI Impact Basin" by Charles E. Rice and Kem C. Kadavy (July, 1991, Vol. 117, No. 7). *J Hydraul Eng-Asce*. [Online] 118(7), 1076–1077. Available from: doi: 10.1061/(ASCE)0733-9429(1992)118:7(1076.2) [Accessed on 24 October 2013].

Colgate, D. (1971). *Hydraulic Model Studies of Corrugated-Metal Pipe Underdrain Energy Dissipators*. [Online] United States Bureau of Reclamation. Report number: REC-ERC-71-10. Available from: http://www.usbr.gov/pmts/hydraulics_lab/reportsdb/reportsearchall.cfm [Accessed 24 June 2013].

Federal Emergency Management Agency (FEMA) (2010) FEMA P-679. *Technical Manual: Outlet Works Energy Dissipators*. [Online] FEMA. Available from: http://www.damsafety.org/media/Documents/DownloadableDocuments/ResourcesByTopic/P679_EnergyDissipators2010.pdf [Accessed on 21 May 2013].

George, R.L. (1979). *T or C Baffled Apron Spillway*. [Online] United States Bureau of Reclamation. Report number: GR-79–2. Available from: http://www.usbr.gov/pmts/hydraulics_lab/reportsdb/reportsearchall.cfm [Accessed 14 June 2013].

Goel, A. & Verma, D.V.S. (2001). Model studies on stilling basins for pipe outlets. *J Irrig Drain Syst*. [Online] 15, 81–91. Available from: doi:10.1023/A:1017989028411 [Accessed 16th May 2013].

Hayes, R.B. (1978). Energy Dissipators: Baffled Apron Drops. In: Aisenbrey, A.J. Jr. (ed.) *Design of Small Canal Structures*. Denver, CO: United States Bureau of Reclamation. pp. 299–308.

Higgins, D.T. (1981). Performance of Baffled Chute with 90° Bend. *J Hydraul Eng-Asce*, 107 (HY4), 419–426.

Jackson, L., Gonzalez-Quesada, P. & Hussein, A. Engineers at Black & Veatch. (Personal communication, 3rd July 2013).

Johnson, M.C. & Dham, R. (2006). Innovative Energy-Dissipating Hood. *J Hydraul Eng-Asce*. [Online] 132(8), 759–764. Available from: doi:10.1061/(ASCE)0733-9429(2006)132:8(759) [Accessed on 31 May 2013].

Johnson, M.C., Sagar, B.T.A & Bergquist, J. (2001). Valves to get out of a fix. *Int Water Power Dam Constr*, July 2001, 22–25.

Lemons, J. Los Angeles Deputy City Engineer. (Letter to R.E. Winter & Associates, Ltd., 22 May 1975). (Provided courtesy of the City of Los Angeles Hydraulics Research Laboratory).

Peterka, A.J. (1978). *Hydraulic Design of Stilling Basins and Energy Dissipators*. [Online] United States Bureau of Reclamation. Engineering monograph number 25. Available from: http://www.usbr.gov/pmts/hydraulics_lab/reportsdb/reportsearchall.cfm [Accessed 16 May 2013].

Rhone, T.J. (1977). Baffled Apron as Spillway Energy Dissipator. *J Hydraul Eng-Asce*, 103 (HY12), 1391–1401.

Rice, C.E. & Kadavy, K.C. (1991). HGL Elevation at Pipe Exit of USBR Type VI Impact Basin. *J Hydraul Eng-Asce*. [Online] 117(7), 929–933. Available from: doi:10.1061/(ASCE)0733-9429(1991)117:7(929) [Accessed on 16 May 2013].

Tennessee Department of Transportation (TDOT) (2010) *TDOT Design Division Drainage Manual*. [Online] Available from: http://www.tdot.state.tn.us/Chief_Engineer/assistant_engineer_design/design/DrainManChap%201-11.htm [Accessed on 3 July 2013].

Thompson, P.L. & Kilgore, R.T. (2006). *Hydraulic Design of Energy Dissipators for Culverts and Channels*. [Online] United States Department of Transportation, Federal Highway Administration. Hydraulic Engineering Circular number 14, 3rd edition. Available from: http://www.fhwa.dot.gov/engineering/hydraulics/pubs/06086/ [Accessed 16 May 2013].

Tiwari, H.L. (2013). Design of Stilling Basin Model with Impact Wall and end Sill. *Res J Recent Sci*. [Online] 2(3), 59–63. Available from: http://www.isca.in/rjrs/V2I3.php [Accessed on 13 May 2013].

United States Army Corps of Engineers (USACE) (1970) HDC(712–1). *Hydraulic Design Criteria: Stone Stability*. [Online] Vicksburg, MS: United States Army Corps of Engineers. Available from: http://chl.erdc.usace.army.mil/hdc [Accessed on 31 May 2013].

United States Army Corps of Engineers (USACE) (1990) EM 1110-2-1603. *Engineering and Design – Hydraulic Design of Spillways: Energy Dissipators*. [Online] Available from: http://www.publications.usace.army.mil/USACEPublications/EngineerManuals.aspx?udt_43544_param_page=4 [Accessed 22 July 2013].

United States Bureau of Reclamation (USBR) (1955). *Progress Report II: Research Study on Stilling Basins, Energy Dissipators, and Associated Appurtenances*. [Online] United States Bureau of Reclamation. Hydraulic laboratory report Hyd-399. Available from: http://www.usbr.gov/pmts/hydraulics_lab/reportsdb/reportsearchall.cfm [Accessed on 16 May 2013].

United States Bureau of Reclamation (USBR) (1958). *Hydraulic Model Studies of the Davis Aqueduct Turnouts at Stations 15.4 and 11.7 Weber Basin Project – Utah*. [Online] United States Bureau of Reclamation. Hydraulic laboratory report Hyd-442. Available from: http://www.usbr.gov/pmts/hydraulics_lab/reportsdb/reportsearchall.cfm [Accessed on 8 July 2013].

United States Bureau of Reclamation (USBR) (1987). *Design of Small Dams*. 3rd Edition. (Section 9.8(i) and Section 10.17(b-c)).

Urban Drainage and Flood Control District (UDFCD) (2008). *Urban Storm Drainage Criteria Manual, Volume 2: Chapter 8-Hydraulic Structures*. [Online] Denver, CO. Available from: http://www.udfcd.org/downloads/down_critmanual_volII.htm [Accessed on 20 May 2013].

Urbonas, B. (burbonas@urbanwatersheds.org) (17 June 2013) *Questions on Storm Drainage Manual*. E-mail to: Bradshaw, R. (riley.bradshaw@aggiemail.usu.edu).

Verma, D.V.S. & Goel, A. (2000). Stilling Basins for Pipe Outlets Using Wedge-Shaped Splitter Block. *J Irrig Drain Eng-Asce*. [Online] 126, 179–184. Available from: doi: 10.1061/(ASCE)0733-9437(2000)126:3(179) [Accessed 16 May 2013].

Verma, D.V.S. & Goel, A. (2003). Development of Efficient Stilling Basins for Pipe Outlets. *J Irrig Drain Eng-Asce*. [Online] 129(3), 194–200. Available from: doi: 10.1061/(ASCE)0733-9437(2003)129:3(194) [Accessed on 16 May 2013].

Young, R.B. (1978). Energy Dissipators: Baffled Outlets. In: Aisenbrey, A.J. Jr. (ed.) *Design of Small Canal Structures*. Denver, CO: United States Bureau of Reclamation. pp. 308–322.

Chapter 7

Energy dissipation: Concluding remarks

H. Chanson
School of Civil Engineering, The University of Queensland, Brisbane, QLD, Australia

ABSTRACT

A key outcome of this monograph is the diversity of energy dissipation designs, the intrinsic complexity of each type of design, and the challenges associated with the magnitude of the energy to be dissipated safely during a flood event. Herein the modelling of energy dissipation structures is developed, together with a discussion of the future in hydraulic structure research and training. It is argued that the technical challenges associated with energy dissipation are not well understood, sometimes understated. The present monograph provides the engineering community with real-world state-of-the-art expertise in energy dissipator design.

7.1 PRESENTATION

At a hydraulic structure, the spill of flood waters involves a number of hydraulic engineering processes, including the transition from subcritical flow in the reservoir to supercritical flow on the chute, free-surface aeration in high-velocity flow, energy dissipation in the stilling structure with a transition from supercritical to subcritical flow motion associated with intense turbulence and further air entrainment. A key outcome of this monograph is the diversity of energy dissipation designs illustrated in Figure 7.1, the intrinsic complexity of each type of design, and the challenges associated with the magnitude of the energy to be dissipated safely during a flood event. Let us consider a small structure carrying 20 m³/s with a drop in elevation of 2 m as seen in Figure 7.1A: the energy loss flux per unit time is 390 kW and the energy loss accounts for nearly 9.4 MW-h per day. For a 155 m high dam spillway discharging a flood flow of 10,000 m³/s (Fig. 7.1B), the kinetic energy of the flow is dissipated at a rate of 15.2 GW (or 15,200,000 kW). Many engineers have never been exposed to the complexity of hydraulic structure designs, to the physical processes taking place and to the structural challenges.

The design of energy dissipators must further consider environmental factors, encompassing the downstream water quality, the effect of supersaturation of dissolved gases on fish and aquatic species, and the impact of atomization caused by flip bucket jets, deflectors and impact dissipators on the eco-environment, which can result in landslides, freezing fog and fog rain.

In the following section, the modelling of energy dissipation structures is developed, together with a discussion of the future in hydraulic structure research and training.

Figure 7.1 Energy dissipation systems. (A) Stilling basin with baffle block downstream of a culvert. (B) Plunge pool at the toe of Sakuma dam spillway (Japan) on 27 November 2008; Dam height: 155.5 m, Design flow: 10,000 m³/s. (C) Baffle chute in southern Manitoba (Canada) (Courtesy of John Rémi).

7.2 OUTLOOKS

7.2.1 Hydraulic modelling of energy dissipators

Design engineers must be able to predict accurately the behaviour of hydraulic structures under the design and non-design conditions including the emergency situations. During the design stages, the engineers need some reliable predictive tools to compare the performances of a range of design options. These tools may be based upon mathematical models, scale models in a laboratory, numerical models and prototype experiences, or a combination of the above. The 20th century saw a widespread use of physical scale models (Novak and Cabelka 1981). Today physical modeling remains a common design tool for complicated hydraulic structures as those shown in Figure 7.2.

The analytical and numerical studies of spillway and energy dissipators are difficult because of the large number of relevant equations and parameters, and the technical challenges to model accurately the highly turbulent flow motion often characterised by intense flow aeration. Numerical modelling is progressing, as discussed in Chapter 4, but the application of a numerical model is too often restricted by the absence of (quality) validation data sets: *"no experimental data means no validation"* (Roache 2009). Some numerical techniques (LES, VOF) may be applied to turbulent flows with large Reynolds numbers, but they lack microscopic resolutions and are not always applicable to air-water flows. Other numerical techniques (DNS) provide a greater level of small-scale details of the turbulent dissipation processes but are limited to small Reynolds numbers. Prototype data are rare because field measurements are extremely difficult and dangerous. Most field data are limited and typically qualitative. Even a few data sets like the Aviemore dam spillway (Cain and Wood 1981) were obtained for a relatively small discharge corresponding to Reynolds numbers less than 5×10^6, one to two orders of magnitude smaller than the design flow conditions of some large spillway systems.

The modelling of energy dissipators remains closely associated with physical modelling using laboratory scale models. A true dynamic similarity is achieved in a geometrically-similar model only if each Π-term has the same value in model and prototype. Gravity effects are important in free-surface flows and a Froude similarity is typically considered (Henderson 1966, Liggett 1994). In the particular case of hydraulic jumps, basic momentum considerations demonstrate the significance of the inflow Froude number (Bélanger 1841, Lighthill 1978) and the selection of the Froude similitude derives implicitly from theoretical considerations (Liggett 1994, Chanson 2012). The Morton number is a constant in most hydraulic modelling studies because both laboratory and full-scale prototype flows use the same fluids, i.e. air and water. In energy dissipators, the turbulent mixing processes involve some viscous dissipation, thus implying a Reynolds similitude. Thus a true dynamic similarity of stilling structures does require achieving identical Froude, Reynolds and Morton numbers in both the prototype and its laboratory scale model. This is impossible to achieve using geometrically-similar models. In practice, the Froude and Morton dynamic similarities are simultaneously employed when air and water are used in the prototype and model. The Reynolds number may be grossly underestimated in laboratory models, leading to viscous-scale effects in small-size models (Novak and Cabelka 1981, Chanson 2009).

Figure 7.2 Spillway systems and operation. (A) Hinze dam spillway Stage 3 (Australia) in operation on 29 January 2013 (Q = 200 m³/s), view from upstream with a stepped chute in foreground, followed by a stilling structure, turning veins, long chute and flip bucket. (B) Intense air entrainment at Three Gorges Project (China) on 20 October 2004 (V = 35 m/s, shutter speed: 1/1,000 s). (C) Sediment deposits in Jiji weir stilling basin (Taiwan) on 11 January 2014 during cleanup. (D) Joe Sippel weir (Australia) on 5 March 2013 after a very large flood event, with a plunge pool/hydraulic jump on the stepped chute in the foreground.

Future studies of energy dissipators may be based upon some 'composite' modelling embedding numerical and physical studies. Composite modelling will not be easy because the modelling of energy dissipators encompasses many challenges including free-surface, air-water flow and turbulence. In parallel, physical modelling must adapt to the needs of composite modelling by delivering high quality experimental data sets encompassing a greater level of details of the boundary conditions and inflow and outflow conditions.

7.2.2   Future hydraulic structure research and training

In many western countries, the decline in dam construction since 1970s has affected large and medium size structures, following changes in government policies, public opinion, and environmental issues. This trend was associated with a shift from innovative design to maintenance, repair, modernisation of an aging dam infrastructure, together with smaller dam construction. The situation led to a diminished opportunity for junior and mid-career hydraulic structure engineers to learn from seasoned experts. In several countries, (too) many senior engineers with large-dam design experience retired and passed away, thus leaving a void in the dam engineering knowledge base (Tullis 2012). While many research results are (usually) available in the literature, a lot of technical expertise and experience, beyond that information, retired with these senior engineers and scholars.

The lack of 'large-dam' engineering experience has become problematic in several countries, sometimes leading to dams and spillways designed overseas, inducing further loss in dam design experience for local engineers, while some hydraulic structure projects might lack the local knowledge of hydrology and hydraulics most necessary to the optimum design. Further, a limited spillway operation means limited engineering experience for the local hydraulic engineers, as mentioned during a recent flood in Australia: "*Obvious matters requiring attention included [...] the lack of any [flood] training exercise*" (Queensland Floods Commission of Inquiry 2011). As a world leader in hydraulic structures, it is the author's opinion that the spillway design expertise in many developed countries has declined for the last 30 years and is no longer leading edge.

The resolution to such a situation is not simple because it requires long-term solutions through sustained investments; there is no '*quick fix*'. Dam engineering is a life-long learning with over 5,000 years in dam engineering (Schnitter 1994) and yet much more can be learned. It is believed that best management practices (BMPs) are not good enough in spillway design, when the state-of-the-art expertise and experience must be the standards. The long-term future of hydraulic structures lies with long-term solutions in terms of teaching, training and research. The transfer of knowledge and the strengthening of engineering fundamentals constitute the prerequisite of hydraulic structure engineering. Junior and senior professionals in dam engineering and hydraulic structures must actively engage in training, higher education (i.e. at postgraduate levels) and information exchanges, combined with active participation to workshops, postgraduate courses, and technical conferences. Dam engineers, together with hydraulic structure researchers, must actively pursue opportunities to share lessons learned and case-study experiences, including presentations in learned societies, specialised symposia, round table, workshops as well as forensic engineering studies and exchanges. At university levels, this must be associated with a drastic reversal in university funding trends in several countries affected by severe tertiary education funding cuts. University students must gain a solid understanding of the fundamental engineering principles (e.g. conservation of mass, momentum, energy) together with good problem-solving skills. These students must develop independent learning skills in hydraulic structures to become effective life-long learners, through literature reading of credible references and physical studies (laboratory, field works) because these do bring some unique life-long experience (Tullis 2012). Professional training and higher education must be complemented by sustained research and development in dam engineering and hydraulic structures. That encompasses industrial research (hydraulics laboratory, physical modelling), innovative research in fundamental hydraulic engineering in industry research centres and university laboratories, and joint industry-university efforts.

7.3   FINAL WORDS

Dams and reservoirs are the most efficient means to provide long-term water reserves and flood protection to our society. A key component for dam safety is the spillway system designed to pass safely flood waters above/below the dam. The design relies upon high level technical expertise and experience in hydrodynamics and hydraulic engineering (conveyance, energy dissipation). The operational challenges are numerous and require a broad, first-hand experience and solid technical expertise. These technical challenges (conveyance, energy dissipation) are not well understood and too often understated. At some recent workshops on hydraulic structures, some world's leading hydraulic engineers shared their concerns: "*our design procedures for major hydraulic structures is often all too optimistic*", "*on many occasions, the expression of one's experience on the extremes of hydraulics behaviour is often countered with skepticism: 'it couldn't happen to my dam, or spillway, or bridge' ...*"; "*there is nothing virtual about hydraulic engineering and hydraulic structures*". It is hoped that the present monograph will provide the engineering community with real-world state-of-the-art expertise in energy dissipator design.

ACKNOWLEDGEMENTS

The author thanks his colleagues, students and all the people who provided relevant informations. The author acknowledges the financial support of the Australian Research Council (Grants DP0878922 & DP120100481).

REFERENCES

Bélanger, J.B. 1841 Notes sur l'hydraulique. ('Notes on hydraulic engineering') *Ecole Royale des Ponts et Chaussées*, Paris, France, session 1841–1842, 223 pages (in French).

Cain, P., and Wood, I.R. (1981) Measurements of self-aerated flow on a spillway. *Journal of Hydraulic Division*, ASCE, Vol. 107, No. HY11, pp. 1425–1444.

Chanson, H. (2009) Turbulent Air-water Flows in Hydraulic Structures: Dynamic Similarity and Scale Effects. *Environmental Fluid Mechanics*, Vol. 9, No. 2, pp. 125–142 (DOI: 10.1007/s10652-008-9078-3).

Chanson, H. (2012). Momentum Considerations in Hydraulic Jumps and Bores. *Journal of Irrigation and Drainage Engineering*, ASCE, Vol. 138, No. 4, pp. 382–385 (DOI 10.1061/(ASCE)IR.1943-4774.0000409).

Henderson, F.M. (1966) *Open channel flow*. MacMillan Company, New York NY, USA.

Liggett, J.A. (1994) *Fluid mechanics*. McGraw-Hill, New York NY, USA.

Lighthill, J. (1978) *Waves in fluids*. Cambridge University Press, Cambridge UK.

Novak, P., and Cabelka, L. (1981) *Models in Hydraulic Engineering. Physical Principles and Design Applications*. Pitman Publ., London, UK.

Queensland Floods Commission of Inquiry (2011) *Interim Report*. Queensland Floods Commission of Inquiry, Brisbane, Australia, Aug., 266 pages.

Roache, P.J. (2009). Perspective: Validation – what does it mean? *Journal of Fluids Engineering*, ASME, Vol. 131, paper 034503 (DOI: 10.1115/1.3077134).

Schnitter, N.J. (1994) *A History of Dams: the Useful Pyramids*. Balkema Publ., Rotterdam, The Netherlands.

Tullis, B.P. (2013) Current and Future Hydraulic Structure Research and Training Needs. *Proceedings of International Workshop on Hydraulic Design of Low-Head Structures*, IAHR, 20–22 Feb., Aachen, Germany, D. Bung and S. Pagliara Editors, Bundesanstalt für Wasserbau, Germany, pp. 3–9.

Subject index